JOURNAL OF
GREEN ENGINEERING

Volume 3, No. 2 (January 2013)

JOURNAL OF GREEN ENGINEERING

Chairperson: Ramjee Prasad, CTIF, Aalborg University, Denmark
Editor-in-Chief: Dina Simunic, University of Zagreb, Croatia

Editorial Board
Luis Kun, Homeland Security, National Defense University, i-College, USA
Dragan Boscovic, Motorola, USA
Panagiotis Demstichas, University of Piraeus, Greece
Afonso Ferreira, CNRS, France
Meir Goldman, Pi-Sheva Technology & Machines Ltd., Israel
Laurent Herault, CEA-LETI, MINATEC, France
Milan Dado, University of Zilina, Slovak Republic
Demetres Kouvatsos, University of Bradford, United Kingdom
Soulla Louca, University of Nicosia, Cyprus
Shingo Ohmori, CTIF-Japan, Japan
Doina Banciu, National Institute for Research and Development in Informatics, Romania
Hrvoje Domitrovic, University of Zagreb, Croatia
Reinhard Pfliegl, Austria Tech-Federal Agency for Technological Measures Ltd., Austria
Fernando Jose da Silva Velez, Universidade da Beira Interior, Portugal
Michel Israel, Medical University, Bulgaria
Sandro Rambaldi, Universita di Bologna, Italy
Debasis Bandyopadhyay, TCS, India

Aims and Scopes
Journal of Green Engineering will publish original, high quality, peer-reviewed research papers and review articles dealing with environmentally safe engineering including their systems. Paper submission is solicited on:

- Theoretical and numerical modeling of environmentally safe electrical engineering devices and systems.
- Simulation of performance of innovative energy supply systems including renewable energy systems, as well as energy harvesting systems.
- Modeling and optimization of human environmentally conscientiousness environment (especially related to electromagnetics and acoustics).
- Modeling and optimization of applications of engineering sciences and technology to medicine and biology.
- Advances in modeling including optimization, product modeling, fault detection and diagnostics, inverse models.
- Advances in software and systems interoperability, validation and calibration techniques. Simulation tools for sustainable environment (especially electromagnetic, and acoustic).
- Experiences on teaching environmentally safe engineering (including applications of engineering sciences and technology to medicine and biology).

All these topics may be addressed from a global scale to a microscopic scale, and for different phases during the life cycle.

JOURNAL OF GREEN ENGINEERING

Volume 3   No. 2   January 2013

Published, sold and distributed by:
River Publishers
P.O. Box 1657
Algade 42
9000 Aalborg
Denmark

Tel.: +45369953197
www.riverpublishers.com

Journal of Green Engineering is published four times a year.
Publication programme, 2012–2013: Volume 3 (4 issues)

ISSN 1904-4720

Green Telecom – An Indian Perspective

Tilak Raj Dua

Global ICT Standardisation Forum for India, Suite 303, 3rd Floor, Tirupati Plaza, Plot No. 11, Pocket-4, Sector 11, Dwarka, New Delhi 110075, India; e-mail: tr.dua@gisfi.org, tilakrajdua@gmail.com

Received 12 September 2011; Accepted: 28 September 2011

Abstract

Eco friendly, energy efficiency, green are the key words in today's market. The reason is not only the health of our planet but also the impact on our business. Combating climate change, securing energy supply and meeting ever increasing energy requirements are the main challenges our society is facing in the present times. There is a vast potential for application of renewable energy options such as wind, bio mass, solar and energy recovery from wastes for meeting partial or total requirement of thermal as well as electrical energy in various industry sectors.

Keywords: Green telecom, India, energy efficiency.

1 Introduction

Combating climate change, securing energy supply and meeting ever increasing energy requirements are the main challenges our society is facing in the present times. There is a vast potential for application of renewable energy options such as wind, bio mass, solar and energy recovery from wastes for meeting partial or total requirement of thermal as well as electrical energy in various industry sectors.

India's high economic growth will lead to an increase in emission of environmentally harmful green house gases that contribute to global warming,

Journal of Green Engineering, Vol. 3, 113–126.

but adopting methods to replace greener or more efficient technologies can help it tap new opportunities as well as get other benefits.

The adoption of greener or more efficient technologies can give India:

- Energy security.
- Inclusive growth.
- Better quality of life.
- Leadership in emerging growth business.

1.1 India's Commitment

India has just announced its commitment to a reduction of 20–25% in carbon intensity from 2005 levels by 2020 through mandatory fuel efficiency standards as announced by Shri Jairam Ramesh, Hon'ble Minister of State for Environment and Forests in the Lok Sabha on December 3, 2009.

1.2 Green Requirement in the Telecom Sector

'Going green' is no more an option for telecom operators. It has become a necessity in a market where margins are nose diving due to tariff wars, denting the profitability of the established service providers. At present the energy expenses (opex) is nearly 25% of the total network operating costs. It is all the more imperative that efficient power management methods/mechanism should be adopted to reduce the operating costs and improving financials

Equipment vendors, tower companies and network services provider are investing heavily in bringing out green products and solutions to cut operators' opex. While a pan-India deployment looks some time away, however incase subsidies and easy availability of renewable energy sources can be ensured it is predicted that up to 20% reduction in energy requirements is achievable in the near future.

In addition, there is awareness building around hybrid and alternative energy sources for the cell sites. The government MNRE has even decided to provide monitory support/subsidy to help operators explore renewable energy and other green technology options.

It is our belief that the telecoms sector must join other industries in going green, to adopt responsible investment strategies, seek out innovative solutions to reduce their carbon emissions, and ultimately established energy security to ensure a long term, sustainable future.

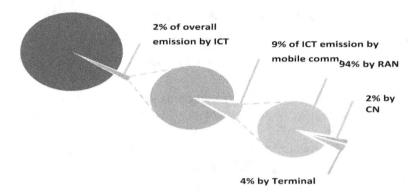

Figure 1 Contribution of ICT towards carbon footprint. Source: Typical Industry Data.

2 Increasing Carbon Footprint – Contribution of the Telecom Industry

A *carbon footprint* is "the total set of greenhouse gases (GHG) emissions caused by an organization, event or product". For simplicity, it is often expressed in terms of the amount of carbon dioxide, or its equivalent of other GHGs, emitted.

Rapid advancements in telecommunications have resulted in unprecedented growth, record infrastructure investments and accelerated service innovation. But as billions of devices access the growing number of networks and platforms on the global information grid, the energy that is used to keep the data flowing and networks buzzing is running out and changing the world's climate.

Though, as an industry, telecommunication generates less pollution and tends to be relatively clean. However, as many other industries, it consumes energy and generates waste.

Greenhouse Gas (GHG) Emissions from the Mobile Industry arise from a number of sources:

- Energy consumed by the network in operation.
- Embedded emissions of the network equipment, for example, emissions associated with the manufacturing and deployment of network equipment.
- Energy consumed by mobile handsets and other devices, when they are manufactured, distributed and used, as well as their embedded emissions.

2.1 Energy Requirements by the Telecom Sector in India

Some of the indicative energy requirement trends in India are:

- Energy related expenditure accounts for nearly 70% of total operating cost per cell site in the rural areas.
- The power requirement of a BTS currently varies from 1300–1500 watts.
- A large percentage of these deployments are still indoor type needing air conditioning.
- Current SLAs (with operators) need shelter temperature to be maintained between a range of 22–25°C.
- Powering systems are largely dependent on grid supply as primary source with diesel generators as stand by sources and storage batteries as secondary sources.
- In view of the current power deficit scenario where load shedding and / or voltage irregularity is at frequent intervals, most of the cellular operators and independent telecom infrastructure providers pre-dominantly use Standby Diesel Generator Sets at their cell sites around the state in order to provide 24×7 uninterrupted cellular mobile services to the end users.

2.2 Concerns

There are the following concerns of the telecom service providers and the environmentalists with regard to the energy requirements:

- Base Stations are still power intensive.
- Grid supplies in rural areas are often erratic and unavailable requiring long runtime of DG sets.
- Diesel generators need regular maintenance.
- Diesel thefts are very prominent – they could be as much as 20% diesel theft.
- Prices of petroleum products are continuously increasing.

3 The Need for Carbon Credit Policy for the Telecom Sector

a. The Kyoto Protocol, 1997, urges all countries to reduce their greenhouse gas emissions by 5% from their 1990 levels by 2012, or pay a price. A carbon credit refers to one tonne of carbon dioxide emissions avoided by the adoption of a certain practice when compared with a business-as-usual (baseline) scenario; it can be sold on the carbon market to a company in the developed world looking to offset excess emissions.

b. The World Bank has built itself a role in this market as a referee, broker and macro-manager of international fund flows. The scheme was entitled Clean Development Mechanism (CDM) – or 'carbon trading' – in 2000. The bank subsequently handed over $10 mn to India's Infrastructure Development Finance Company to fund clean projects that would generate saleable carbon credits.

c. The concept of carbon trading arrived in India in 2002, and since then India has developed an attractive CDM portfolio with a market share close to 12%.

d. *Carbon Credit Policy for Indian Telecom Sector*: The following are some of the suggestions:

  – In a bid to promote environment-friendly telecom infrastructure, it is suggested that the Government may propose to give carbon credits to operators for using eco-friendly fuels to power their exchanges and mobile base stations.

  – Government/regulator may recommend giving financial incentives in terms of lower revenue share to operators deploying non-conventional sources of energy such as solar and wind energy wherever possible, as the operational cost to provide backup power supply in case regular electric supply is erratic, is very high. Moreover, since the use of generators for long hours results in pollution, there is a need to encourage use of non-conventional sources of energy.

  – If a service provider uses, say, solar energy to energize base stations, the company may be considered for certain incentives. Taking a cue from similar concepts being used internationally to reduce pollution. These incentives should be based on the International Best practices.

  – The telecom players are now investing in a host of telecom products that are increasingly contributing towards both building a greener tomorrow and helping the country's economy, while showing its customers that being environment-friendly is profitable. The energy solutions companies are innovating on technologies that can reduce GHG emissions drastically, in process making the unit or the project eligible for carbon credits in large volumes.

- With over 60,000 sites of telecom operates using the green energy solutions, more than 840960 tonnes of carbon is being saved each year. Credits that can be earned from this is quite large.

4 Methods and Options to Reduce the Carbon Footprint by the ICT Industry in India

The mitigation of carbon footprints through the development of alternative projects, such as solar or wind energy or reforestation, represents one way of reducing a carbon footprint.

Methods to reduce the carbon foot print by ICT industry in India calls for a combination of incentives and subsidies, including market and fiscal mechanisms to help environment management by industry and people in their day to day working. Environment education and awareness is also critically important in this context. Other options are:

i. Energy consumption is a significant ingredient in running and maintaining telecom networks. Reducing the carbon footprint should be a adopted as a good practice for the telecom sector, inclusive of service providers and the associated industries of the sector, particularly, the telecom equipment manufacturers.

ii. Government should provide Incentives for the development and use of Alternative energy such as solar or wind energy

iii. The guidelines need to be in place for recycling of waste materials.

iv. The procedure for installation of new infrastructure should be aligned with the environmental policies.

v. Adequate thrust should be on Environment education and awareness creation.

vi. Investment in new technology which contains less hazardous material and is, thus, easy to recycle.

vii. Introduction of energy efficient technologies.

5 Options for Environment Friendly Alternate Energy Sources

Many telecom companies are now exploring multiple sources of renewable energy, like solar, wind, biofuels, etc. A lot many are going with choices like green wireless networking equipment.

Other countries have started getting huge outputs from their stakes in renewable energy. By 2010, five solar thermal electricity generators in the Australian desert will produce enough electricity for a million homes.

Exploring alternative sources of energy is not only imperative now but is also seen as a viable option that can help with a cleaner and greener environment and also generate job opportunities in the rural part of the country.

In India, more than 80,000 villages do not even have a grid electricity pole anywhere near. Supplying power to these areas still remains a challenge for telecom companies.

Adapting alternative sources of energy for powering BTS sites is essential. It is estimated that 118,000 renewable energy base stations could save up to 2.5 billion litres of diesel a year and cut annual carbon emissions by up to 6.3 million tons (Industry Estimates).

Various sources of energy that can power the BTS are:

(a) Solar Energy

    – Solar-DG Hybrid
    – Solar-wind Hybrid

(b) Wind-DG Hybrid
(c) Biomass Gasifier
(d) Biofuels blending with diesel

(a) *Solar Energy*

Solar Energy is the *most matured* of all such technologies. The main advantages are:

- Clean & green.
- No moving parts – minimal maintenance cost.
- Easier to manage.

Where can we leverage Solar Energy?

- Use Solar Energy as primary source in a 'NO Grid' situation.
- Ideal for rural area with little chance of shadow effect.
- Zero carbon emissions.
- Less possibility of site outage as site running on solar power during day time.

Suggested configuration:

a. Solar-DG Hybrid
b. Solar-Wind Hybrid

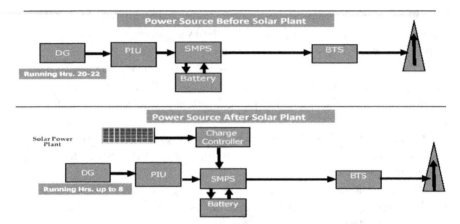

Figure 2 Schematics of solar deployment for Telecom.

(b) *Wind-Based System*
This provides *free, clean & green* energy.

- Advanced systems are widely available.
- Smaller systems can be mounted on existing radio-masts, reducing costs.
- Horizontal wind turbines are more efficient.
- Systems available with low "cut-in" speeds of 2.4 m/sec.

Some challenges are:

- Site-selection must be carefully done for deployment of wind turbines, (ISO-820 wind maps must be studied before deploying wind turbines)
- Wind velocity is often erratic. Thus we need a very efficient charge controller and a sink for excess power
- Sink for excess power can be a tube well for example

(c) *Biomass Gasifier*
Biomass gasification is basically a conversion of solid fuels (wood/wood-waste, agricultural residues, etc.) into a combustible gas mixture normally called Producer Gas.

The process is typically used for various biomass materials and it involves partial combustion of such biomass partial combustion process occurs when air supply (O_2) is less than adequate for the combustion of biomass to be completed

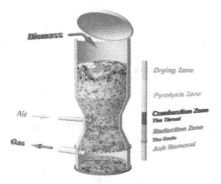

Figure 3

The DG which is part of the previous hybrid solution can be made to use biomass instead of costly and polluting fossil fuel. It reduces the dependence on diesel.

Some concerns are:

- Availability of raw material on continuous basis at identified locations.
- Requirement of large covered storage space for biomass storage.
- Dedicated manpower 24 × 7 for day to day operation.
- Safe disposal of ash/residue.

(d) *Biodiesels*

Biofuel is a fuel oil made mainly from organic vegetable oils or animal fats. The production process is called tran sesterification. Every vegetable oil molecule consists of fatty acids and glycerin, the trans esterification process separates the fatty acids from glycerin by adding an alcohol and a catalyst. This results in biodiesel with properties similar to fossil diesel. The use of biodiesel in a conventional diesel engine results in substantial reduction of unburned hydrocarbons, carbon monoxide, and particulate matter compared to emissions from diesel fuel.

Biodiesel is an alternative fuel to regular conventional diesel (fossil fuel) and can be produced from vegetable oils, acid oils, fatty acids, fish oil, mutton tallow and used cooking oils. Few of the non-edible vegetable oils like Pongamia and Jatropha also can be used to produce biodiesel.

This fuel meets all the international standards namely ASTM/EN/BIS standards and can be used as 100% or in various blends with regular diesel. It is accepted by global players, such as Mercedes Benz, Volkswagen, Cater Pillar, Cummins, Skoda, etc.

Biodiesel in India:

- India is the 6th largest diesel consumer in the world.
- Indian railways alone consume 4 billion liters per annum.
- Projected annual growth of diesel usage in India is 5.8%.
- India consumed around 52 million metric tons of diesel in the last year.
- The estimated demand for biodiesel is between 3.15 million metric tons to 12.60 million metric tons by 2010 (subject B5 to B20).
- Around 130 million acres of wasteland available in India.
- Can produce around 130 million tons of biodiesel through biodiesel plantations in the wastelands, considering 40% land suitable for biodiesel plantation.
- At present Jatropa and Pongamia are the plantations considered as biodiesel plantation.
- The Government of India is encouraging biodiesel plantations by extending various subsidies to farmers.

6 VAS Green

There have been recent initiatives by the mobile operators wherein they have promoted a slew of value added services like mobile newspaper, receive e-bills on mobile, mpayments and transactions, issue e-tickets and boarding passes, thereby saving tonnes of paper each day.

The operators have long been talking about the business sense in adopting green measures. It is one industry that pays hefty energy bills. The initiative to educate subscribers on the use of mobile applications that can help subscribers cut down their carbon footprint is an encouraging step towards the long journey

Industry experts say VAS usage may not contribute in a big way, but definitely it can make a significant impact in the long run. A whole variety of value added services are on the offer that help you mind your carbon footprints. Their carbon emission calculators, alarms for mobile and what not, but services like mobile newspaper and e-tickets on mobile definitely make a direct contribution.

Most of the Indian telecom service providers have on and off been making some efforts to encourage subscribers to adopt green practices. For example, on Diwali operators made offers where a subscriber opting to get e-bills would stand a chance of getting his house painted for free. In India, their prepaid and postpaid ratio is skewed. Though a majority of the subscribers are prepaid, the number of postpaid customers is around 6 to 7%. Thus, there is a huge amount of saving that can be done by receiving bills online rather than getting prints of bills.

7 Contribution of the Industry to Reduce the Carbon Footprint

The industry can contribute by having specific focus groups and programs in areas such as:

- Awareness building and knowledge dissemination – Indian ICT industry needs to be educated about impact of ICT on CO_2 emissions – both the negative impact and how ICT can be used to reduce emissions.
- Energy labeling programs – There is a need to enhance existing Energy Labeling programs in India.
- Metrics programs – There is a need to come out with metrics and performance indicators for energy efficiency of all relevant ICT activity in India. This includes harmonization of global best practices and guidelines for India. Also, there is scope to develop new guidelines for areas in which there are no global standards or best practices so far – for example energy efficiency metrics for Radio Base Stations.
- Code of conduct – There is a need for establishing a Code of Conduct for Indian ICT operations similar to the European Code of Conduct for achieving time bound targets of energy efficiency.

7.1 Global Efforts in Green ICT

i. Global e-Sustainability Initiative

- GeSI is a non-profit organization, headquartered in Brussels, Belgium that brings together ICT companies, industry associations and NGOs to further the cause of sustainable economy using innovative use of ICT.
- GeSI's activities include development of standards, methodologies, best practices and promotion of good conduct in the areas of Climate Change, Energy Efficiency, E-waste and Supply Chains.

ii. The Green Grid

- The Green Grid is global consortium of over 180 organizations worldwide.

- The Green Grid focuses on development of promotion of energy efficiency in data centers and enterprise computing environments.
- It also promotes the adoption of energy efficient standards, processes, measurement methods and technologies.

iii. Climate Savers Computing Initiative

- The Climate Savers is a non-profit organization started in 2007 by Google and Intel.
- The focus of this group is on energy efficient PC and servers.

iv. US Energy Star Program

- Energy Star is an initiative of the US Environmental Protection Agency and Department of Energy.
- The Energy Star program for computers aims to generate awareness of energy saving capabilities and accelerate the market penetration of more energy- efficient technologies.

v. Smart Grid Interoperability Panel

- The Smart Grid Interoperability Panel (SGIP) is a membership-based organization created by the US National Institute of Standards and Technology (NIST).
- NIST is responsible for coordinating the development of and publishing a framework, including protocols and model standards, to achieve interoperability of Smart Grid devices and systems.

vi. Initiatives of the European Commission

- EC has acknowledged the importance of use of innovative ICT-based technologies for achieving a low carbon emission targets in a cost effective manner.
- The Commission is promoting R&D projects for developing ICT tools to improve energy efficiency

8 Conclusions

- India is well positioned to take off for alternative energy revolution; however organized efforts are yet to pick up stream.
- The telecom sector must join other industries in going green, to adopt responsible investment strategies, seek out innovative solutions to reduce their carbon emissions, and ultimately established energy security to ensure a long term, sustainable future.

- Government support in terms of subsidy is required to bolster up the Green Energy usage in the telecom sector.
- The Government should consider USOF support to encourage operators to opt for green energy and bioffuel as an alternative for powering BTS.
- More efforts are required to educate the industry on the need for cleaner fuel, its environmental importance and the socio-economic benefits of biofuel for the rural areas.
- The green VAS initiatives of the operators can be a milestone in educating subscribers on the use of mobile applications that can help subscribers cut down their carbon footprint is an encouraging step towards the long journey.

It is yet to be seen that 'green' does not remain just a marketing differentiator, but actually becomes a compulsory ingredient of all the equipments, products that are rolled out by telecom operators in India.

References

[1] GISFI Standardisation Workshops, www.gisfi.org.
[2] Department of Telecommunications – Ministry of Communications & IT, Committee Report, www.dot.gov.in.
[3] FICCI, Study Paper, http://www.ficci.com/studies&surveys.asp.
[4] Green Peace Workshop, http://members.greenpeace.org/event/view/4446/.
[5] Ministry of New & Renewable Energy Sources, Report, http://www.mnre.gov.in/.

Biography

T.R. Dua is a B.Sc. Engineering graduate with a diploma in Business Management and Export Marketing. He has more than 35 years of experience in Telecommunication in various fields, such as product development, the introduction of new technologies, technical collaborations, joint ventures, telecom regulations, and spectrum management (spectrum auction, pricing, and allocation). He was Director of Shyam Telecom Ltd., Director of Bharti Airtel Ltd., and Deputy Director General of COAI.

Dr. Dua is a member of: ITU – APT, PTC India, Fellow Institution of Engineers (India), Fellow Institute of Electronics and Telecom Engineers, Computer Society of India, Indian Science Congress, Optical Society of India, 3GPP, WWRF, GSMA, Task Group on Spectrum – Digital dividend, and the National Working Group on Spectrum WP5D/NWGS. Furthermore, he is currently on the Governing Council of "Global ICT Standardization Forum of India".

In 2010 he was awarded the prestigious "Evangelist Mobile Infrastructure Award". He has published and/or presented numerous papers on Spectrum optimization Techniques, Transition from IPV4 to IPV6, Green Telecom, Mobile Money Transfer, VOIP Security & Mitigation Techniques, EMF radiation/alleged hazards.

Currently his interests include: EMI/EMC, satellite, microwave, and spectrum optimization.

On the Design of an Optimal Hybrid Energy System for Base Transceiver Stations

Panagiotis D. Diamantoulakis and George K. Karagiannidis

Department of Electrical and Computer Engineering, Aristotle University of Thessaloniki, Thessaloniki, Greece; e-mail: {padiaman, geokarag}@auth.gr

Received 30 November 2012; Accepted 28 December 2012

Abstract

The reduction of energy consumption, operation costs and CO_2 emissions at the Base Transceiver Stations (BTSs) is a major consideration in wireless telecommunications networks, while the utilization of alternative energy sources, such as solar or wind, having emerged as an attractive solution with numerous advantages. Nevertheless, the installation of BTSs with renewable energy, induces specific disadvantages such as the relatively higher costs and the high dependency on weather conditions. To this end, the deployment of hybrid BTSs and the optimal compromise between conventional and alternative energy sources is a very challenging problem with immense importance. In this paper, we propose a hybrid solar-wind-diesel/electricity grid system, which can efficiently feed the load of a BTS. In contrast to previous works, the seasonal effect on the BTS's electricity consumption is taken into account via considering (i) the radio transmitter and receiver operation strategy, based on the cells's traffic load, and (ii) the passive cooling. The main objective of the present work is the techno-economical optimization of the proposed hybrid system, via the development of a time-step simulation model, which takes into account the loss of load probability (LOLP) and levelized annual cost (LAC). Finally, considering the case-study of a BTS installed in the Greek island of Kea, it is shown that a combination of photo-voltaic, wind, diesel generators, batteries and electricity grid, for a grid-connected BTS, is the most cost-effective solution.

Journal of Green Engineering, Vol. 3, 127–146.

Keywords: Hybrid energy system, Base Transceiver Station (BTS), energy consumption, energy optimization.

1 Introduction

It is estimated that the telecom industry consumes about 1% of the global energy consumption of the planet [1, 2]. That is, the equivalent energy consumption of 15 million US homes and also the equivalent CO_2 emissions of 29 million cars. Over 90% of the wireless networks energy consumption is part of the operator's operating expenses. There are approximately 4 million installed Base Transceivers Stations (BTSs) in the world today. A BTS of a wireless communications network consumes 100 watts of electricity to produce only 1.2 Watts of transmitted radio signals. From a system efficiency perspective (output/input power), this translates into an energy efficiency of 1.2% [1]. In a typical BTS, radio equipment and cooling are the two major sections where the highest energy savings potential resides. To the best of our knowledge, the following techniques are proposed in the literaure [1] to reduce the power consumption of BTS:

- moving the RF converters and power amplifiers from the base of the station to the top of the tower close to the antenna and connecting them via fiber cables;
- turning radio transmitters and receivers off when the call traffic goes down (ECO Mode);
- passive cooling;
- advanced climate control for air conditioners;
- DC Power Sustem ECO Mode;
- use of higher efficiency rectifiers.

Renewable energy [3], such as wind and solar, is also used to reduce the operation costs in BTSs. These energy sources are intermittent, naturally available, environmental friendly and can be used to provide virtually free energy. Solar systems are a mature technology, used to power some remote BTSs for many years, replacing the expensive to run diesel generators. Hybrid solar-wind systems use two renewable energy sources, improving the system efficiency and reducing the energy storage requirements [4]. A solar-wind hybrid power generation system for remote BTSs is also proposed in [5]. However, the main problem of the renewable energy installations is that the generation of electricity cannot be fully forecasted and may not follow the trend of the actual energy demand [6]. In order to select an optimum combination for a hybrid system to meet the load demand, evaluations must be carried out on the basis of power reliability and system life-cycle cost.

Recently, several simulations have been performed in order to optimize hybrid energy systems and to fulfill the energy demands of a BTS. Hashimoto et al. [7] described a stand-alone hybrid power system, which is consisted of wind generators and photovoltaic modules for a BTS, comparing the produced energy during the worst month of the year with the consumed energy. In [8, 9] the software Homer (Hybrid optimizing model for electric renewables) is used to size a hybrid energy system for BTS [4]. Homer is a time-step simulator using hourly load and environmental data inputs for renewable energy system assessment. However, the limitation of this software is that algorithms and calculations are not visible or accessible. Ekren et al. [10] used ARENA 10.0, a commercial simulation software, to meet the electric power consumption of the global system for mobile communications (GSM) base station. One of the main objectives of Ekren et al. [10] is to show the use of the response surface methology (RSM) in size optimization of an autonomous PV/wind integrated hybrid energy system with battery storage. Furthermore, the same authors, in [11], developed a Simulated Annealing (SA) algorithm to optimize the size of a PV/wind integrated hybrid system with battery storage. They show that the SA algorithms give better results than the RSM. Furthermore, in [12], ARENA 12.0 was used to size a PV/wind integrated hybrid energy system with battery storage under various loads and unit cost of auxiliary energy sources. The optimum results were confirmed using loss of load probability and autonomy analysis. Finally, Hongxing et al. [13] proposed an optimal design model for designing hybrid solar-wind system employing battery banks for calculating the system optimum configurations and ensuring that the systems's annualized cost is minimized, while satisfying the custom required loss of power supply probability. The proposed method has been applied to design a hybrid system to supply power for a telecommunication relay station. The previous studies ignore the seasonal effect on the BTSs electricity consumption or they focus primarily on remote BTSs.

In this paper, we propose a hybrid solar-wind-batteries-diesel/electric grid system to reduce the operation costs in TBSs and an appropriate sizing model to evaluate them. The development of the time-step simulation model is based on the loss of load probability and levelized annual cost. The recommended algorithm is appropriate to size properly a hybrid system in order to meet all or a part of sustained load demands of a BTS and to evaluate the CO_2 emissions. The appropriate operation of a BTS supplied by hybrid system is described, modeled and simulated. The considered consumption model is a typical BTS, as presented by [14].

The contribution of this paper can be summarized as follows:

1. The proposed model is suitable both for remote and grid-connected BTSs.
2. The strategies of passive cooling and ECO Mode, which are responsible for the seasonal effect on the BTS's electricity consumption, are not ignored. The final BTS's load is determined during the time of simulation, via considering the indoor and outdoor temperature for the determination of the power consumption of the cooling equipments. In order to model the aforementioned techniques, we considered the separation of the total BTS's load in two parts:

   (a) the DC part, which includes the radio equipment, the feeder and the antenna, and
   (b) the AC part, which includes the cooling equipments.

3. An alternative use of cooling equipment is proposed. This new method is consisted of full operation of cooling equipments when there is a surplus of energy produced by the solar-wind system and limited operation when the energy provided by the solar-wind-batteries system almost meets the BTS's consumption.
4. Finally, the proposed algorithm is used to cost several configurations of hybrid system for a BTS, located in the Greek island of Kea. It is shown that a combination of photo-voltaic generators, wind generators, batteries and diesel generator or grid is the most cost-effective solution.

The rest of the paper is organized as follows. In Section 2, we introduce the hybrid system configuration. The model for BTS load is given in Section 3, while the power management is formulated in Section 4. The algorithm for the techno-economic analysis is presented in Section 5 and the simulation results are given in Section 6. Finally, the paper is concluded in Section 7.

2 The Hybrid System Configuration

2.1 Wind Turbine

The main factor that determines the power output of a wind turbine is the wind speed. Therefore, choosing a suitable model is very important for wind turbine power simulations. The most simplified model to simulate the power output, P_w of a wind turbine can be described by the following equation [15]:

$$P_w(u) = \begin{cases} 0, u \leq U_c \\ P_r \frac{u-U_c}{U_r-U_c}, U_c < u < U_r \\ P_r, U_r \leq u \leq U_f \\ 0, u > U_f \end{cases} \tag{1}$$

where P_r is the rated power, U_c is the cut-in wind speed, U_r is the rated wind speed and U_f is the cut-off wind speed.

Wind speed varies with height, while wind data of different sites are measured at different levels. The wind speed, u, corresponding to a height, H, is given by the wind power law for the case of known value of wind speed, u_0, for the specific height H_0, as

$$u = u_0 \left(\frac{H}{H_0}\right)^\xi, \tag{2}$$

where ξ is the power law exponent. The one-seventh is a good reference number for flat surfaces where a BTS is usually located, far away from tall trees or buildings. Here, H and H_0 denote the height of the turbine and measuring above the ground, respectively.

2.2 Photo-Voltaic Generator

There are three main factors that determines the power output of a PV generator, namely the PV cell material, the cell temperature and the solar radiation incident on the PV modules. An appropriate model to simulate the power output of a PV generator is described by [15]

$$P_p = n_p E_p G_t, \tag{3}$$

where n_p represents the PV generator efficiency, E_p (in m^2) the area of the PV generator and G_t (in W/m^2) the solar radiation that incidents on the PV module. The PV generator efficiency is given by

$$n_p = n_m n_e [1 - b(T_c - T_r)], \tag{4}$$

where n_m is the reference module efficiency, n_e is the power conditioning efficiency, b is the generator efficiency temperature coefficient, T_r is the reference cell temperature and T_c is the cell temperature given by

$$T_c = T_a(t) + \frac{T_n - 20}{800} G_t, \tag{5}$$

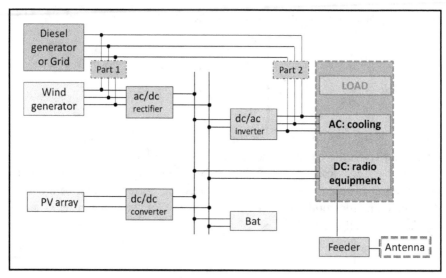

Figure 1 System configuration.

with T_a being the ambient temperature and T_n the nominal cell operating temperature.

The solar radiation incident G_t on a PV module is a function of time, orientation and slope angle. However, the solar radiation is usually measured only at a horizontal plane. Thus, a technique to transfer the radiation measured at horizontal plane to the desired slope angle β of the PV module is needed. Such a technique is described in detail in [16].

2.3 Diesel Generator – Grid

The majority of the BTSs already has a diesel generator, which can also be used as a backup to the hybrid system, reducing the installed size of the described wind/PV/battery system. At the same time, grid connection could be used as a back up too, when possible. Moreover, the grid should be used as a storage system, by giving the surplus of energy to the grid or taking from it when there is a shortfall. The aforementioned technique leads to great financial benefits, such as the nulling of the batteries required. However, the unique disadvantage associated with this technique is the increase of the CO_2 emissions.

The operation of the diesel generator can be described as $P_n = n_t P_d$, where P_d is the nominal power of the diesel generator and $n_t = n_0 n_g$ [17].

The factor n_0 is used to show that the total efficiency of the diesel generator depends on the load factor. Diesel generators work more efficiently when they are operated at 70–90% of full load. Since the proposed hybrid energy system will include batteries, it is recommended that they charge when the diesel generator is already on.

By means of the algorithm we describe in Section 5, the CO_2 emissions of every solution can be computed. Obviously on a system that relies only on green sources of energy the emissions are null. But when diesel generator or grid is used, CO_2 emissions per year, F, can be defined as

$$F = E_m X_i, \qquad (6)$$

where E_m is the energy that the BTS takes from diesel generator, while in the case of grid connection, it is the energy that the BTS takes from it without to return it later, considering that BTS can sell energy back to the grid. The parameter X_i (in $[kgCO_2/KWh]$), is the carbon emissions of the not green source of energy (diesel generator/grid) that occasionally feeds the BTS. However, in the case of grid connection, the increase of CO_2 emissions may be balanced with an equivalent corresponding decrease, when the surplus of energy given to the grid is equal to the energy taken previously from it.

2.4 Battery System

According to the previous section, the cooling power (AC load) is not only a function of the power that consumes the radio equipment (DC power), but it also depends on the operation of passive cooling. For this reason, it has to be considered as a different part of consumption during the simulation. Note that during discharging process, the energy that the batteries can offer is around 20% of their nominal capacity, and the maximum depth of discharge is around 30–50%, according to the specifications of the manufacturers [16]. Furthermore, before the power reaches the air conditioner, it passes trough an inverter. In fact, it is not a simple inverter, but its study is not on the purpose of this article. We only care of its efficiency and cost.

When $K < 0$ and the capacity in the time t is higher than the minimum allowable, the batteries are in discharging process and the storage capacity in the instant $t + \Delta t$ is given by

$$C_b(t + \Delta t) = \begin{cases} C_b(t) + \frac{K}{n_d}, \quad \frac{|K|}{n_d} < \frac{C_m}{5}\Delta t \\[2ex] C_b(t) - \frac{C_m}{5}\Delta t, \quad \frac{|K|}{n_d} \geq \frac{C_m}{5}\Delta t. \end{cases} \qquad (7)$$

When $K > 0$ and the batteries are not full, then they are in charging process and the storage capacity is given by

$$C_b(t + \Delta t) = \begin{cases} C_b(t) + Kn_h, \ Kn_h < \frac{C_m}{5}\Delta t \\ \\ C_b(t) + \frac{C_m}{5}\Delta t, \ Kn_h \geq \frac{C_m}{5}\Delta t, \end{cases} \tag{8}$$

where $C_b(t + \Delta t)$ is the available battery bank capacity at time $t + \Delta t$, $C_b(t)$ is the available battery bank capacity at time t, n_h is the battery charging efficiency, n_d is the battery discharging efficiency, C_m is the nominal batteries' capacity and Δt is the time-step (in hours) of the simulation. For the parameter K, it holds that

$$K = (P_{p1} + P_w)n_r + \frac{P_{p2} - A_C}{n_i} + P_p n_c - D_C, \tag{9}$$

where variable A_C is used to describe the AC part of consumption, D_C is the DC part, n_i is the efficiency of the inverter, n_r is the rectifier efficiency and n_c is the converter efficiency. Note that P_{p1} is the power that goes from diesel generator or grid to DC or to batteries trough Part 1 and P_{p1} is the power that goes from diesel generator or grid to AC trough Part 2. The two parts of consumption, DC and AC, are described in more detail in the following section.

In the case of discharging, the offered energy (P_b) by the batteries at time $t + \Delta t$ can be expressed as

$$P_b = (C_b(t + \Delta t) - C_b(t))n_d. \tag{10}$$

3 BTS Load

As mentioned previously, the two major consumers of energy on a BTS are the radio and the cooling equipment. In more details, the radio equipment consumes about 68% of the total energy to produce the RF power, the air conditioner is responsible for the 30% of the consumption, 1% is losses on the feeder and only 1% goes to the final signal [1].

3.1 DC Load

In order to define the DC load of a BTS, namely the consumption of RF, feeder and antenna, a strategy called ECO Mode has to be taken into account during the technical and economic assessment of the hybrid system. This

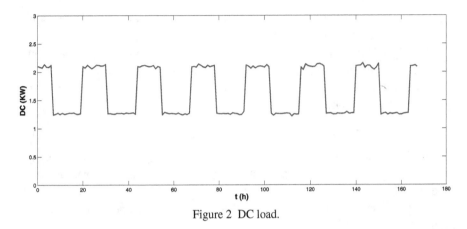

Figure 2 DC load.

strategy consists of turning radio transmitters and receivers off when the call traffic goes down. This typically happens during the night (see Figure 2), reducing the power consumption of DC up to 40% [1].

3.2 AC Load

A strategy that reduces the power consumption of a BTS is the operation of the air conditioner at higher temperature. It has been calculated that by allowing a wider fluctuation between 31 and 26°C, the total cooling cost which can be saved is about 14% [1].

In this paper, an appropriate modification of the aforementioned strategy is adopted, in order to better taking the advantage of the output power of the hybrid energy system. Particularly, we suggest that (i) when the produced energy of the BTS hybrid system exceeds consumption and the batteries are full, the surplus of energy should be used to cool the BTS up to a lower temperature, and (ii) when the produced energy is not enough, consumed energy on the air condition could be eliminated. The only restriction of this strategy is that the temperature level must not exceed the maximum desirable one. Using the two aforementioned strategies, wind or PV generators of lower nominal power can meet the electric power consumption, reducing the total costs of the hybrid system and making the approach of hybrid energy system for BTSs more attractive. Furthermore, it needs to be emphasized that the cooling requirements do not exclusively depend on the power which is consumed on the radio equipment. Especially in the BTSs where the technique of passive cooling is used, the cooling requirements also depend on the weather conditions, such as temperature.

Hourly AC load, which is mainly consisted of the consumption of the air-condition, is given by $A_C(t) = Q_L(t)/n_a$, where n_a is the efficiency of the air-condition and $Q_L(t)$ is the total cooling load per hour. For the parameter Q_L, it holds [18] that

$$Q_L(t) = Q_I(t) + Q_A(t) + Q_E(t) + Q_S(t), \tag{11}$$

where Q_I is the heat production from the RF, which is taken as a percentage of DC consumption, Q_A the air leakage, Q_E the heat transfer through envelope of BTS and Q_S the solar radiation transferred trough envelope, which in this paper is considered to be equal to zero. The parameters Q_A, Q_E are defined as

$$Q_A = GdC_p(T_0 - T_r) \tag{12}$$

and

$$Q_E = KF(T_0 - T_r), \tag{13}$$

assuming that G is the air leakage flow rate through envelope, d density of air, C_p the specific heat capacity of air, K the envelope heat transfer coefficient, F the total envelope's surface area, while T_0 and T_r are hourly outdoor and indoor temperatures respectively [18].

Hence, cooling load of BTS depends on the outside temperature and during the winter, which is fairly considered as the worst period for a hybrid wind/PV system, the final load of is lower. Considering the previous statement, the solution of hybrid power system for BTSs can be regarded much more feasible.

If a fluctuation of temperature is allowed, the indoor temperature of BTS can be given by

$$T_{in}(t + \Delta t) = \frac{Q(t + \Delta t)}{m_a C_p} + T_{in}(t), \tag{14}$$

where

$$Q(t) = Q_L(t) - Q_I(t) - Q_E(t) - Q_A(t) \tag{15}$$

and T_{in} is the indoor temperature.

4 Power Management

There are some basic principles that have to be taken into account for an efficient power management during the operation of a BTS. The load of BTS must use the available green sources of energy as much as possible. The diesel generator and grid provide power on priority to AC. If a fluctuation of the temperature is not unacceptable, the use of a diesel generator or grid should be avoided. When the diesel generator is already on and its load factor is lower than 90%, the diesel generator should be used to charge the batteries, if the batteries are not full.

A_C is the air condition consumption that the internal cabinet temperature will remain constant and equal to a default temperature. A_{Cn} is the down limit of air condition consumption. Air condition consumption cannot be considered smaller than A_{Cn}, because that would provoke an increase of indoor temperature larger than the desirable one. So the essential air condition consumption, A_{Cn} is defined by the upper limit, T_{upl}, and under limit, T_{unl}, of the indoor temperature. The exact operation of the hybrid system is described in Figure 3.

5 Techno-Economic Analysis

5.1 Description of the Algorithm that Controls the Suitability of the Solution

Reliability is a very important factor for a hybrid system that feeds a BTS. On the other hand, the generation of energy by wind turbines and PV generators is a stochastic process and thus, time to time simulation is needed to check which are the components of the hybrid system and the characteristics of the BTS in order to secure the reliability of the system. Reliability is controlled by the number of loss of load probability. Loss of load probability (LOLP) is the probability – the percentage of time – that the hybrid system does not satisfy the load of BTS. So LOLP can be defined as

$$\text{LOLP} = \frac{\tau}{T}, \tag{16}$$

where T is the number of hours in this study with weather data input and

$$\tau = \sum_{t=0}^{T} \mathcal{D}\left[(P_{p1} + P_w)n_r + \frac{(P_{p2} - A_{Cn})}{n_i} + P_p n_c + P_b - D_C < 0\right]. \tag{17}$$

In (17), $\mathcal{D}[x]$ denotes the time duration where the x event is true.

A high value of LOLP means that system is not reliable. A solution is unsuitable when LOLP is larger than an upper limit, which is defined by L.

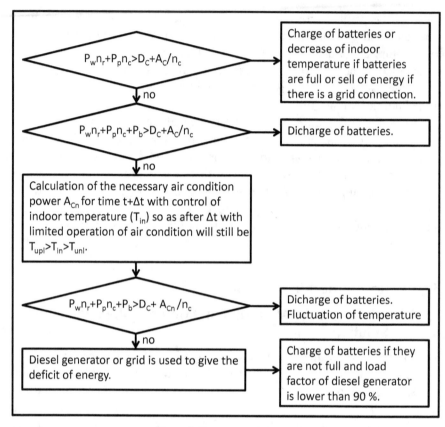

Figure 3 Power management.

5.2 Economic Evaluation of the Solution

The cost of a hybrid system depends on the following factors: the nominal power of wind and PV generators, the number of batteries, the life time of the hybrid system, the usage of diesel generator or grid, the replacement cost of batteries batteries, the maintenance cost of the hybrid system. From all the suitable combinations of parts of hybrid system, optimum is the one with the lower cost. The present value of the total cost [13] can be given by

$$V_p = V_c + V_m + V_i + V_r, \tag{18}$$

where V_c is the initial capital cost of the system, which depends on the nominal power of wind turbines (P_r), the number of PV generators (N_p), the nominal power of diesel generator (P_d), the number of batteries (N_b) and their

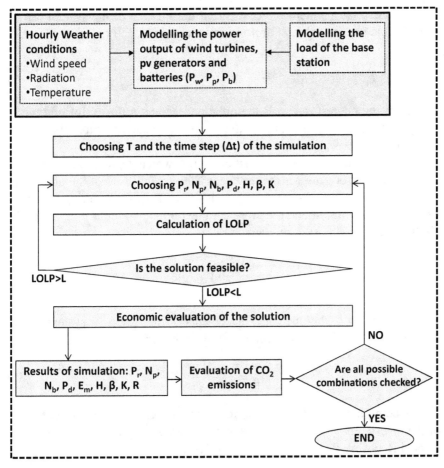

Figure 4 Simulation process.

costs, V_m is the present value of the total maintenance cost of the system, V_i is the present value of the cost of energy produced by diesel generator or taken from grid and V_r the present value of total replacement cost of batteries and during the life time of hybrid system. The total maintenance cost of hybrid system in the first year can be defined as

$$M_t = M_p + M_w + M_b, \qquad (19)$$

where M_p, M_w, M_b is the maintenance cost of PV generators, wind turbines and batteries in the first year respectively. The maintenance cost of system every next year is higher because of the annual inflation rate. The present value of the total maintenance cost for all the lifetime of the system can be given by the present value of uniform inflation series factor, f.

$$f_s = \frac{1 - (\frac{1+a}{1+i})^n}{i - a},$$ (20)

where i is the interest rate, a is the inflation rate and n the useful lifetime of the system. So the present value of the total maintenance cost for all the years is given as $V_m = M_t f_s$.

Assuming that B is the total cost of energy, which is provided either from diesel generator or grid, for the first year, the present value of the aforementioned cost of energy for all the years is $V_i = B f_s$. Considering that only batteries need replacement, the present value of the total replacement cost for all the lifetime of the system is given by

$$V_r = \sum_{j=1}^{y} (V_{cb} f_p),$$ (21)

where y is the total number of replacements of the batteries' system for all the years and V_{cb} the initial capital cost of the batteries.

$$f_p = \frac{1}{(1 + i)^x},$$ (22)

with x being the year of every one replacement. Knowing the present value of the investment, V_p, the levelized annual cost, R, can be calculated by the Capital Recovery Factor, f_r, as $R = V_p f_r$, where

$$f_r = \frac{i}{1 - (1 + i)^{-n}}.$$ (23)

6 Results and Discussion

In this section, we present results of the techno-economic analysis of the hybrid system, for the use case of a BTS in the Greek island of Kea. In case study, we consider that the useful lifetime of the system is 25 years, while the upper limit of LOLP, L, must be equal to 0. Real hourly weather data for the year of 2011 were used for the simulation, which are obtained from Meteorological Institute of National Observatory of Athens. We assume that wind speed, solar radiation and temperature are constant during the time step.

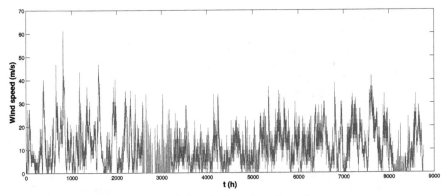

Figure 5 Hourly data of wind speed.

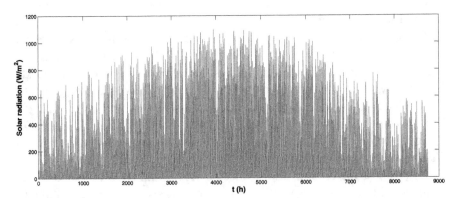

Figure 6 Hourly data of solar radiation.

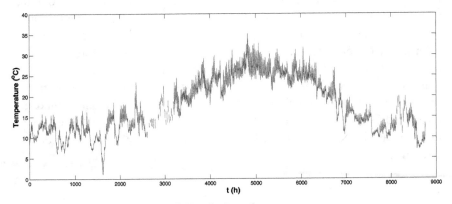

Figure 7 Hourly data of temperature.

Table 1 Levelized annual cost for autonomus BTS (normalized to the cost of the exclusive use of diesel generator).

System Configuration	LAC (EUR)
PV-Wind turbines-Batteries-Diesel generator	0.472
Wind turbines-Batteries-Diesel generator	0.585
PV-Batteries-Diesel generator	0.792
Diesel generator	1

The considered hourly wind speed, solar radiation and temperature profiles are shown in Figures 5, 6 and 7.

Due to the high availability of wind and solar potential in Kea, the installation of systems based on renewable sources of energy shows to be a promising solution for BTSs. In order to clarify the previous statement, both the cases of autonomous and grid-connected BTS are simulated and analyzed. As shown at Tables 1 and 2, a combination of PV generators, wind turbines, batteries and diesel generator or grid, for grid connected BTSs, is the most cost-effective solution both for autonomous and grid connected BTSs. Note that we get the used values for PV, wind wind generators, diesel, etc from the current market. Particularly, for the case of remote BTS, the autonomous diesel generator system leads to 1 levelised (normalized) annual system cost (R), while PV-Wind turbines-Batteries-Diesel generator system leads to R equal to 0.472, using the diesel generator only for 20% of the total time. This translates into 52.8% reduction of operation costs of system and 80% reduction of the CO_2 emissions. For the case of grid-connected BTS, utilizing a hybrid PV-wind turbines-grid system leads to 12% reduction of R and 64% reduction of CO_2 emissions, compared to the case of the conventional grid-connected system. In this study we considered that BTSs can sell energy back to the grid at half the price as they buy it. In Tables 1 and 2, we also show that a configuration of system based on the combination of the two renewable sources of energy, solar and wind energy, is the most cost effective solution. Moreover, as we have observed, stand-alone PV/wind energy systems, without a diesel generator or electric grid for back up, are not reliable, as they cannot guarantee the zero loss of load probability. At both the cases of autonomous and grid connected BTS, the load of BTS mostly relies on the production of wind turbines, as shown in Figure 8. However, the existence of PV generators is cost effective, because these two green sources of energy are complimentary.

Table 2 Levelized annual cost for grid connected BTS (normalized to the cost of the exclusive use of grid).

System Configuration	LAC (EUR)
PV-Wind turbines-Grid	0.879
Wind turbines-Grid	0.907
PV-Grid	0.915
Grid	1

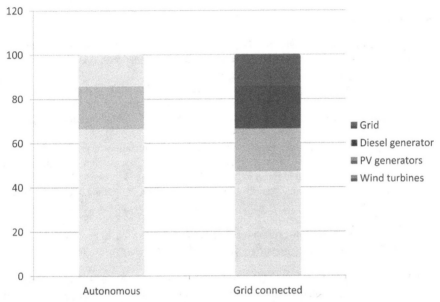

Figure 8 Distribution of energy to different systems (%).

7 Conclusions

The operational costs of BTSs can be effectively reduced through the utilization of renewable energy sources. Nevertheless, despite their significant benefits, such as their environmental impact and their substantial availability, they also induce specific disadvantages such as the high initial capital cost and the dependency on weather conditions. In this paper, we presented a hybrid system, which uses renewable energy sources (solar and wind energy), diesel power and the electric grid. This system has been optimized for minimizing the operational costs of BTS, while promising high reliability. Moreover, the proposed time-step simulation is appropriate for all BTSs, either connected to the grid or not. The seasonal effect on the BTS's electric energy consumption, as well as the smarter operation of air-conditioner and diesel generator are taken into account for the technical and economical analysis. According to numerical results, for the use case of the Greek island of Kea, we confirmed that hybrid energy system is a promising, cost-effective option for both remote and grid-connected BTSs, via reducing remarkably the total annualized cost of energy system and CO_2 emissions.

References

[1] Steve N. Roy. Energy logic: A road map to reducing energy consumption in telecommunications networks. In Proceedings of IEEE 30th International Telecommunications Energy Conference, San Diego, CA, 2008.

[2] W. Vereecken, W. Van Heddeghem, M. Deruyck, B. Puype, B. Lannoo, W. Joseph, D. Colle, L. Martens, and P. Demeester. Power consumption in telecommunication networks: Overview and reduction strategies. IEEE Communications Magazine, 49(6):62–69, June 2011.

[3] Pragya Nema, R.K. Nema, and Saroj Rangnekar. Minimization of green house gases emission by using hybrid energy system for telephony base station site application. Renewable and Sustainable Energy Reviews, 14(6):1635–1639, August 2010.

[4] Wei Zhou, Chengzhi Lou, Zhongshi Lia, Lin Lu, and Hongxing Yang. Current status of research on optimum sizing of stand-alone hybrid solarwind power generation systems. Applied Energy, 87(2):380–389, February 2010.

[5] A. Naikodi. Solar-wind hybrid power for rural Indian cell sites. In Proceedings of 2010 IEEE International Energy Conference and Exhibition (EnergyCon), Manama, pp. 69–72, 2011.

[6] H. Dagdougui, R. Minciardi, A. Ouammi, M. Robba, and R. Sacile. A dynamic decision model for the real-time control of hybrid renewable energy production systems. IEEE Systems Journal, 4(3):323–333, September 2010.

[7] Shigeo Hashimoto, Toshiaki Yachi, and Tatsuo Tani. A new stand-alone hybrid power system with wind generator and photovoltaic modules for a radio base station. In Proceedings of 26th Annual International Telecommunications Energy Conference, pp. 254–259, 2004.

[8] P. Nema, S. Rangnekar, and R.K. Nema. Pre-feasibility study of PV-solar/wind hybrid energy system for GSM type mobile telephony base station in Central India. In Proceedings of 2nd International Conference on Computer and Automation Engineering (ICCAE), Singapore, 2010.

[9] Niankun Zhang, Zechang Sun, Jie Zhang, Tiancai Ma, and Jiang Wang. Optimal design for stand-alone wind/solar hybrid power system. In Proceedings of International Conference on Electronics, Communications and Control (ICECC), Ningbo, pp. 4415–4418, 2011.

[10] Orhan Ekren and Banu Yetkin Ekren. Size optimization of a PV/wind hybrid energy conversion system with battery storage using response surface methodology. Applied Energy, 85(11):1086–1101, November 2008.

[11] Orhan Ekren and Banu Yetkin Ekren. Size optimization of a PV/wind hybrid energy conversion system with battery storage using simulated annealing. Applied Energy, 87(2):592–598, February 2010.

[12] Banu Yetkin Ekren and Orhan Ekren. Simulation based size optimization of a PV/wind hybrid energy conversion system with battery storage under various load and auxiliary energy conditions. Applied Energy, 86(9):1387–1394, September 2009.

[13] Hongxing Yang, Zhou Wei, and Lou Chengzhi. Optimal design and techno-economic analysis of a hybrid solarwind power generation system. Solar Energy, 86(2):163–169, February 2009.

[14] Ericsson, Sustainable energy use in mobile communications. White Paper, Ericsson, August 2007.

[15] S. Diaf, M. Belhamel, M. Haddadi, and A. Louche. Technical and economic assessment of hybrid photovoltaic/wind system with battery storage in Corsica island. Energy Policy, 36(2):743–754, February 2008.

[16] Hongxing Yang, Wei Zhou, Lin Lu, and Zhaohong Fang. Optimal sizing method for stand-alone hybrid solarwind system with LPSP technology by using genetic algorithm. Solar Energy, 82(4):354–367, April 2008.

[17] P. Nema, R.K. Nemab, and S. Rangnekara. A current and future state of art development of hybrid energy system using wind and PV-solar: A review. Renewable and Sustainable Energy Reviews, 13(8):2096–2103, October 2009.

[18] Rang Tu, Xiao-Hua Liu, Zhen Li, and Yi Jiang. Energy performance analysis on telecommunication base station. Energy and Buildings, 43(2/3):315–325, February/March 2011.

Biographies

Panagiotis D. Diamantoulakis was born in Thessaloniki, Greece in January 1989. He received his Diploma in 2012 in Electrical and Computer Engineering, from the Aristotle University of Thessaloniki, Greece.

Currently he is working toward his Ph.D. degree at the same university. His research interests include smart power systems and green communications.

George K. Karagiannidis was born in Pithagorion, Samos Island, Greece. He received the University Diploma (5 years) and Ph.D degree, both in electrical and computer engineering from the University of Patras, in 1987 and 1999, respectively. From 2000 to 2004, he was a Senior Researcher at the Institute for Space Applications and Remote Sensing, National Observatory of Athens, Greece. In June 2004, he joined the faculty of Aristotle University of Thessaloniki, Greece where he is currently Associate Professor (four-level academic rank system) of Digital Communications Systems at the Electrical and Computer Engineering Department and Head of the Telecommunications Systems and Networks Lab.

His research interests are in the broad area of digital communications systems with emphasis on communications theory, energy efficient MIMO and cooperative communications, cognitive radio and optical wireless communications.

He is the author or co-author of more than 170 technical papers published in scientific journals and presented at international conferences. He is also author of the Greek edition of a book on *Telecommunications Systems* and co-author of the book *Advanced Wireless Communications Systems* (Cambridge Publications, 2012). He is co-recipient of the Best Paper Award of the Wireless Communications Symposium (WCS) in the IEEE International Conference on Communications (ICC'07), Glasgow, U.K., June 2007.

Dr. Karagiannidis has been a member of Technical Program Committees for several IEEE conferences such as ICC, GLOBECOM, VTC, etc. In the past he was Editor for *Fading Channels and Diversity of the IEEE Transactions on Communications*, Senior Editor of *IEEE Communications Letters* and Editor of the *EURASIP Journal of Wireless Communications & Networks*. He was Lead Guest Editor of the special issue on "Optical wireless communications" of the *IEEE Journal on Selected Areas in Communications* and Guest Editor of the special issue on "Large-scale multiple antenna wireless systems".

Since January 2012, he is the Editor-in Chief of *IEEE Communications Letters*.

OFDM AF FG Relaying as an Energy Efficient Solution for the Next Generation Mobile Cellular Systems

E. Kocan and M. Pejanovic-Djurisic

Faculty of Electrical Engineering, University of Montenegro, Podgorica, Montenegro; e-mail: {enisk, milica}@ac.me

Received 28 November 2012; Accepted 29 December 2012

Abstract

We analyzed the bit error rate (BER) and ergodic capacity performance improvement of dual-hop OFDM amplify and forward (AF) relay system with fixed gain (FG) at relay station (R), all attained through ordered subcarrier mapping (SCM) at R. In previous works we showed that in the region of small values of average signal-to-noise ratios (SNRs) on hops best-to-best SCM (BTB SCM) scheme should be applied, while in the region of medium and high values of average SNRs on hops, BER performance improvement is achieved by implementing best-to-worst SCM (BTW SCM) scheme. In this paper we analyzed a new solution, where in the assumed relay system with BTB SCM, one or more subcarriers having the lowest SNRs on each hop are omitted. Furthermore, we examined power balance of the system implementing SCM through comparison of BER performance expressed as a function of power transmitted per subcarrier, with BER performance of the ordinary OFDM AF FG relay system, as well as with the performance in case of direct transmission between source of information (S) and destination terminal (D). OFDM based relay systems are already included in IMT-Advanced standards, and all the obtained BER and ergodic capacity results have confirmed that OFDM AF FG relay system with appropriate SCM scheme implemented at

Journal of Green Engineering, Vol. 3, 147–165.

R station represents very interesting and energy efficient solution for the next generation mobile cellular systems.

Keywords: OFDM, amplify-and-forward, relay, fixed gain, ordered subcarrier mapping, BER, ergodic capacity, energy efficiency.

1 Introduction

The expansion of wireless communication systems over the last decade was enabled through implementation of appropriate technical solutions, which have succeeded to meet customer's demands. In order to achieve the required high data rates and quality of service level, the next generation of cellular systems, IMT-Advanced systems, will incorporate, among other advanced technical solutions, OFDM (Orthogonal Frequency Division Multiplexing) based relay (R) stations, for the system capacity enhancement and coverage area extension, [1]. The accepted relay solution assumes dual-hop scenario, where the source terminal (S) cannot achieve direct communication with the destination terminal (D). Regenerative R stations, that decode the received signal and again re-encode it before forwarding toward destination (D) terminal, are already included in the accepted standards for IMT-Advanced systems, while non-regenerative R stations, that amplify and forward (AF) the received signal, are expected to become a part of the next IMT-Advanced specifications. R station implementing AF technique may amplify the received signal with the fixed gain (FG), or with the variable gain (VG), depending on its possibility to estimate the channel between the source of information (S) and R. Non-regenerative relay systems are less complex than regenerative systems, and they introduce shorter latency. Analyses presented in this paper consider the OFDM non-regenerative (AF) relay system applying FG at the R station.

Although the standards for both IMT-Advanced systems, LTE-Advanced and Wireless MAN-Advanced, have been accepted, there is still ongoing intensive research activity on the OFDM based relay systems, all with the goal to further improve their performance, energy efficiency or to optimize their functions. One of the methods for capacity enhancement and/or bit error rate (BER) improvement, which is at the same time energy more efficient, is implementation of ordered subcarrier mapping (SCM) at the R station, where the subcarriers from the S-R link (first hop) are mapped to appropriate subcarriers on the R-D link (second hop), all in accordance with their instantaneous signal-to-noise ratios (SNRs) [2–8]. The concept of subcarrier mapping (SCM) at the R station was introduced in [2], and shortly afterwards was discussed in [3] and [4]. In these papers it was proved that the capacity of dual-hop OFDM based relay system can be maximized if the subcarriers from the first hop are ordered according to their instantaneous SNRs, and then mapped to corresponding subcarriers on the second hop, which are also ordered in accordance with their instantaneous SNRs. This SCM scheme is denoted as Best-to-Best SCM (BTB SCM). However, when BER performance of the OFDM AF based relay systems is considered, it was shown that the BTB SCM scheme minimizes the BER results only in the region of small SNRs [5]. On the other side, for the medium and high SNR values, the SCM scheme denoted as Best-to-Worst SCM (BTW SCM) is proposed for implementation in order to minimize BER performance [5, 6]. BTW SCM scheme assumes that the subcarriers from the first hop, which are increasingly ordered according to their SNRs, are mapped to the decreasingly ordered subcarriers on the second hop. We derived closed-form BER performances of differentially phase shift keying (DPSK) and binary phase shift keying (BPSK) modulated OFDM AF FG relay systems with SCM in [6, 7], thus analytically proving the optimality of the previously described proposal. We have also examined analytically ergodic capacity performances of OFDM AF FG relay systems with SCM in [7]. However, some issues regarding energy efficiency, and possibility for implementation just one SCM scheme for both capacity enhancement and BER performance improvement, remained opened.

Thus, in order to examine the power balance of SCM solution, in this paper we compare BER performance of differentially phase shift keying (DPSK) modulated OFDM AF FG relay system implementing SCM, expressed as a function of power transmitted per subcarrier, with BER performance of ordinary OFDM AF FG relay system and with the case of direct OFDM transmission between S and D. Additionally, we analyzed if omitting one or more subcarriers having the lowest SNRs on both hops may sufficiently improve BER performance of OFDM AF FG relay system with BTB SCM, to attain or outperform, in all SNR regions, BER performance of the same system implementing BTW SCM. We proved in [8] that this kind of solution in OFDM AF VG relay system makes BTB SCM optimal SCM scheme for BER improvement for all SNR values on both hops. Also, in this paper we give capacity analyses of OFDM AF FG relay system with BTB SCM, through derivation of both upper and lower bounds of ergodic capacity, and present the level of capacity improvement in comparison with the same system without SCM. The paper is organized as follows. Section 2 describes

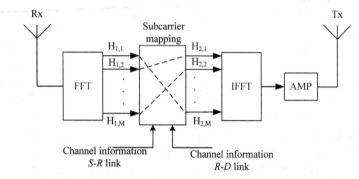

Figure 1 Block scheme of OFDM AF FG relay station with SCM.

the analyzed system model. In Section 3, analytical derivation of BER and ergodic capacity performance is given. The obtained results are presented in Section 4. Finally, in Section 5 we present some concluding remarks.

2 System Model

We considered the same relaying system as the one presented in [7]. It assumes a dual-hop OFDM FG AF relay system, in which three communication terminals, all equipped with single antenna are involved in communication process. There is no direct link between S and D, and R operates in half-duplex mode. The basic block scheme of a relay terminal is presented in Figure 1. R first performs OFDM demodulation through FFT (Fast Fourier Transformation). Then, the subcarriers from the first hop are mapped to the appropriate subcarriers on the second hop in accordance to their instantaneous SNRs. It is assumed that the SCM block knows channel state information on both S-R and R-D links, The next processing step is getting back the signal into time domain through IFFT (Inverse Fast Fourier Transformation) block. The obtained OFDM signal is amplified by Furthermore, the gain G, and then transmitted toward destination terminal. In order to correctly demodulate the received signal D has to know subcarrier permutation function performed at R.

The post-FFT signal on the i-th subcarrier, received at the relay station, is given by:

$$Y_{R,i} = X_{1,i} H_{1,i} + N_{1,i}, \quad 1 \leq i \leq M, \tag{1}$$

where M is the total number of subcarriers and X_i is the data symbol sent by source on the i-th subcarrier. $N_{1,i}$ represents additive white Gausssian noise on the i-th subcarrier with variance $\mathbf{E}(|N_{1,i}|^2) = N_{01}$, with $\mathbf{E}(\cdot)$ denoting the expectation operator. Assuming that the SCM function $v(i)$ performs mapping of the i-th subcarrier from the first hop to the k-th subcarrier on the second hop, the frequency domain signal at D can be written as

$$
\begin{aligned}
Y_{D,k} &= GH_{2,k}Y_{R,v(i)} + N_{2,k} \\
&= GH_{2,k}H_{1,i}X_i + GH_{2,k}N_{1,i} + N_{2,k}, \quad 1 \le k \le M
\end{aligned}
\tag{2}
$$

where $H_{2,k}$ denotes the k-th subcarrier transfer function on the second hop. $N_{2,k}$ is the additive white Gaussian noise at the destination on the k-th subcarrier, with variance $\mathbf{E}(|N_{2,k}|^2) = N_{02}$.

The fadings in the S-R and R-D channnels are assumed to be independent and identically distributed (i.i.d.) among the subcarriers. Moreover, we assume Rayleigh fading in each subcarrier, so that the PDF and the CDF (Cumulative Distributive Function) of the SNR in each of the S-R subchannels is given by $f_{SR}(x) = \lambda_{SR} \exp(-\lambda_{SR}x)$ and $F_{SR}(x) = 1 - \exp(-\lambda_{SR}x)$, while the corresponding PDF and CDF of the SNR in each of the R-D subchannels are given by $f_{RD}(x) = \lambda_{RD} \exp(-\lambda_{RD}x)$ and $F_{RD}(x) = 1 - \exp(-\lambda_{RD}x)$, respectively. $\lambda_{SR} = 1/\bar{\gamma}_{SR}$ and $\lambda_{RD} = 1/\bar{\gamma}_{RD}$ denote the inverse of the average SNRs on the S-R and R-D links, respectively. Using (2) and the described gain, the end-to-end SNR on the k-th subcarrier can be presented as

$$
\gamma_{k,\text{end}} = \frac{\gamma_{i\text{SR}}\gamma_{k,\text{RD}}}{\gamma_{k\text{RD}} + \rho},
\tag{3}
$$

where ρ is the coefficient that depends through the gain G as $\rho = \epsilon_R/(G^2 N_{01}) \cdot \gamma_{i,\text{SR}}$ and $\gamma_{k,\text{RD}}$ represent instantaneous SNRs on the i-th subcarrier of the S-R link and k-th subcarrier of the R-D link, respectively, while ϵ_R is the symbol energy transmitted by R.

3 Performance Analysis

In the following part we use the earlier obtained results on probability density function (PDF) and moment generating function (MGF) of SNR for the k-th subcarrier at D [6, 7], which are necessary tools for conduction of BER and ergodic capacity performance analysis.

3.1 BER Performance Analysis

We have conducted BER performance analysis for the dual-hop DPSK modulated OFDM AF FG relay system with SCM, in the case of Rayleigh fading channels on both hops in [6]. Using the ordered statistics, we first derived PDF of the increasingly ordered random variables having exponential distribution, and through the similar approach, PDF of SNR for decreasingly ordered exponentially distributed random variables. These PDF functions actually correspond to PDF of the SNRs of the ordered subcarriers from the first hop or from the second hop in the considered scenario with Rayleigh fading statistics. Using this, after some mathematical transformations, we derived PDF of SNR for the k-th subcarrier at D for the systems implementing BTB SCM and BTW SCM. However, for the case of DPSK modulation implemented, it is more convenient to use MGF based approach for BER performance analysis, as in that case BER for the k-th subcarrier at D is obtained through [9]:

$$P_{b,k} = 0.5 \cdot \mathcal{M}_{\gamma_{k,\text{end}}}(1), \tag{4}$$

with $\mathcal{M}_{\gamma_{k,\text{end}}}(\cdot)$ denoting MGF of SNR function. BER for the analyzed OFDM AF FG relay system with SCM is derived through averaging (4) over all M subcarriers:

$$P_b = \frac{1}{M} \sum_{k=1}^{M} P_{b,k}, \tag{5}$$

Thus, we derived MGF of SNR at the k-th subcarrier at D for the relay systems implementing both analyzed SCM schemes. In the scenario with BTB SCM scheme at the R station, MGF of SNR for the k-th subcarrier at D is obtained as [6]:

$$\mathcal{M}_{\gamma_{k,\text{end}}}(s) = \frac{1}{\bar{\gamma}_{\text{SR}}} \sum_{j=0}^{k-1} \sum_{i=0}^{k-1} \frac{\alpha_j \alpha_i}{T_j(s)}$$

$$\times \left[\frac{1}{\beta_i} + \exp\left(\frac{\rho A_{j,i}}{T_j(s)} \right) E_1 \left(\frac{\rho A_{j,i}}{T_j(s)} \right) \left(\frac{\rho}{\bar{\gamma}_{\text{RD}}} - \frac{\rho A_{j,i}}{\beta_i T_j(s)} \right) \right]. \tag{6}$$

where $E_1(\cdot)$ represents the exponential integral function defined in [10, (5.1.1)]. The coefficients α_i and β_i are given through:

$$\alpha_i = (-1)^i M \binom{M-1}{k-1} \binom{k-1}{i} \quad \text{and} \quad \beta_i = i + M - k + 1 \tag{7}$$

In (7), (:) denotes binomial coefficients. The coefficients $A_{j,i}$ and $T_j(s)$ are introduced in (6) for the easier representation of this relation, and they are equal to:

$$A_{j,i} = \beta_j \beta_i / \bar{\gamma}_{\text{SR}} \bar{\gamma}_{\text{RD}} \quad \text{and} \quad T_j(s) = s + \beta_j / \bar{\gamma}_{\text{SR}}. \tag{8}$$

Using the same approach, MGF of SNR for the k-th subcarrier at D in the case of BTW SCM scheme implemented at R is derived as [6]:

$$
\mathcal{M}_{\gamma_{k,\text{end}}}(s) = \frac{1}{\bar{\gamma}_{\text{SR}}} \sum_{j=0}^{k-1} \sum_{i=0}^{M-k} \frac{\alpha_j \delta_i}{T_j(s)}
$$

$$
\times \left[\frac{1}{\varepsilon_i} + \exp\left(\frac{\rho B_{j,i}}{T_j(s)} \right) E_1\left(\frac{\rho B_{j,i}}{T_j(s)} \right) \left(\frac{\rho}{\bar{\gamma}_{\text{RD}}} - \frac{\rho B_{j,i}}{\varepsilon_i T_j(s)} \right) \right]. \quad (9)
$$

In (9) the coefficients δ_i and ε_i are equal to

$$
\delta_i = (-1)^i M \binom{M-1}{k-1} \binom{M-k}{i} \quad \text{and} \quad \varepsilon_i = i+k, \quad (10)
$$

while $B_{j,i}$ can be written as

$$
B_{j,i} = \beta_j \varepsilon_i / \bar{\gamma}_{\text{SR}} \bar{\gamma}_{RD}. \quad (11)
$$

By substituting (6) or (9) in (4) and then in relation (5), BER performance of DPSK modulated OFDM AF FG relay system implementing BTB SCM or BTW SCM is obtained.

As we already have mentioned, it is known that OFDM AF FG relay system implementing BTW SCM outperforms the system with BTB SCM in terms of achievable BER performance, in the regions of medium and high values of average SNRs on both hops. The same holds for the OFDM AF VG relay systems implementing SCM [9]. However, in [9] we have shown that in the OFDM AF VG relay system with BTB SCM, omitting just the worst subcarriers on both hops, i.e. the subcarriers with the lowest SNRs, makes the system with BTB SCM optimal solution for BER performance improvement in all the SNR regions. Thus, we wanted to examine if we do not use the subcarriers with the lowest SNRs on both hops in the OFDM AF FG relay system with BTB SCM may improve enough BER performance of this system to attain and prevail BER performance of OFDM AF FG system with BTW SCM, in all the SNR regions. Analytically, it means that in the relation (5) summation starts from $k = 2$ for the system with BTB SCM.

In order to examine if the additional signal processing in systems with SCM can be justified considering the issue of power consumption, the obtained BER results should be presented as a function of power transmitted per subcarrier in the whole system, P_T. We assumed equal power allocation among the S and R station, i.e. $P_S = P_R = P_T/2$ and among all the subcarriers. Now, the average SNRs on S-R and R-D links can be written as

$\bar{\gamma}_{SR} = A_1 P_S$ and $\bar{\gamma}_{RD} = A_2 P_R$, respectively, where A_1 and A_2 include para-meters as the antenna gains, path loss, noise power and similar. For example, if using Friis propagation model, A_i $(i = 1, 2)$, can be written in the form

$$A_i = \frac{G_{t,i} G_{r,i} \lambda^2}{(4\pi)^2 d_i^\alpha L N_{0i}}, \qquad (12)$$

where $G_{t,i}$ is the transmitter antenna gain on the i-th hop, $G_{r,i}$ is the receiver antenna gain, λ is the wavelength, d_i is the distance between the transmitter and receiver on the i-th hop, L is the system loss factor, $\alpha = 2$ for free space and $3 < \alpha < 4$ in urban environment, while N_{0i} is the noise variance at the i-th hop. Without loss of the generality, we took that the transmitter antenna gains at the S and R are equal, $G_{t,1} = G_{t,2}$, and the receiver antenna gains at the R and D are also equal, $G_{r,1} = G_{r,2}$, as well as that the noise variances at the R and D are the same, $N_{01} = N_{02}$. Moreover, we assumed that in the case of relayed transmission, S, R and D are placed on a straight line, and that all the links are affected by the same shadowing environment. The average SNR at D in the case of direct transmission can be written as , where for this simplified propagation model, by taking $\alpha = 3$, A_{eq} is related to A_1 and A_2 through

$$A_{eq} = \frac{A_2}{(1 + (A_2/A_1)^{1/3})^3}. \qquad (13)$$

For the case of direct OFDM transmission in Rayleigh fading environment, MGF of SNR for the k-th subcarrier has the form

$$\mathcal{M}_{\gamma_{k,end}}(s) = \frac{1}{1 + s\bar{\gamma}_{SD}} \qquad (14)$$

Substituting (14) into (4) and then in (5), BER of DPSK modulated OFDM system is obtained.

3.2 Ergodic Capacity Analysis

It is proven that BTB SCM scheme maximizes the achievable ergodic capa-city of OFDM based relay systems [2–4]. Thus, in this part we will analyze only the ergodic capacity performance of the OFDM AF FG relay system with BTB SCM. For the ergodic capacity analysis it is necessary to know PDF of SNR for the k-th subcarrier at D. In the case of BTB SCM scheme applied at R, this function has the form as given in [7]:

$$
f^{\text{BTB}}_{\gamma_k,\text{end}}(x) = \frac{2}{\bar{\gamma}_{\text{SR}}} \sum_{j=0}^{k-1} \sum_{i=0}^{k-1} \alpha_j \alpha_i \exp\left(-\beta_j \frac{x}{\bar{\gamma}_{\text{SR}}}\right)
$$
$$
\times \left[\sqrt{\frac{\rho \beta_j x}{\beta_i \bar{\gamma}_{\text{SR}} \bar{\gamma}_{\text{RD}}}} K_1\left(2\sqrt{\frac{\rho \beta_j \beta_i x}{\bar{\gamma}_{\text{SR}} \bar{\gamma}_{\text{RD}}}}\right) + \frac{\rho}{\bar{\gamma}_{\text{RD}}} K_0\left(2\sqrt{\frac{\rho \beta_j \beta_i x}{\bar{\gamma}_{\text{SR}} \bar{\gamma}_{\text{RD}}}}\right) \right],
$$

(15)

where $K_0(\cdot)$ and $K_1(\cdot)$ are zero and first order modified Bessel functions of the second kind defined in [10, (9.6.21), (9.6.22)]. The presence of these functions in the expression for the PDF of SNR for the k-th subcarrier at D prevail the possibility for finding closed-form solution for the ergodic capacity of this system. However, taking into account the concave form of the logarithmic function, and the Jensen's inequality, we found upper bound of the k-th subcarrier ergodic capacity through

$$
C_k = 0.5 \cdot \mathbf{E}(\log_2(1 + \gamma_{k,\text{end}})) \leq 0.5 \log_2(1 + \mathbf{E}(\gamma_{k,\text{end}})).
$$

(16)

where $\mathbf{E}(\cdot)$ denotes the expectation operator, and factor 0.5 appears due to symbol transmission in two time slots (half-duplex relay). Closed-form expression is derived for $\mathbf{E}(\gamma_{k,\text{end}})$ in [7] as

$$
\mathbf{E}(\gamma_{k,\text{end}}) = \sum_{j=0}^{k-1} \sum_{i=0}^{k-1} \frac{\alpha_j \alpha_i}{\beta_j^2 \sqrt{\rho A_{j,i}}} \exp\left(\frac{\rho A_{i,j} \bar{\gamma}_{\text{SR}}}{2\beta_j}\right)
$$
$$
\times \left[2\sqrt{\frac{\rho \beta_j}{\beta_i} \frac{\bar{\gamma}_{\text{SR}}}{\bar{\gamma}_{\text{RD}}}} W_{-2,1/2}\left(\frac{\rho A_{j,i} \bar{\gamma}_{\text{SR}}}{\beta_j}\right) \right.
$$
$$
\left. + \frac{\rho \sqrt{\beta_j} \bar{\gamma}_{\text{SR}}}{\bar{\gamma}_{\text{RD}}} W_{-3/2,0}\left(\frac{\rho A_{j,i} \bar{\gamma}_{\text{SR}}}{\beta_j}\right) \right].
$$

(17)

In (17) $W_{x,y}(\cdot)$ denotes the Whittaker function defined in [10, (13.1.32)]. The average ergodic capacity of the OFDM AF FG system with SCP is derived through averaging ergodic capacities per subcarrier, over all M subcarriers:

$$
C = \frac{1}{M} \sum_{k=1}^{M} C_k.
$$

(18)

In order to attain more complete insight in ergodic capacity performance of OFDM AF FG relay system with SCM, we also derived a tight lower bound of its achievable ergodic capacity. Using the expression of end-to-end SNR for the k-th subcarrier (3), ergodic capacity of the k-th subcarrier at D can be split in the following way [11]:

$$C_k = I_A - I_B = I_x - I_y - I_B. \tag{19}$$

I_x is derived in closed-form as [11]:

$$I_x = \frac{4}{\ln(2)} \sqrt{\frac{\rho}{\bar{\gamma}_{SR}\bar{\gamma}_{RD}}} \sum_{i=0}^{k-1} \alpha_i \exp\left(\frac{\beta_i}{\bar{\gamma}_{SR}}\right) \sum_{j=0}^{k-1} \alpha_j \sqrt{\frac{1}{\beta_j \beta_i}} S_{-2,1}\left(2\sqrt{\frac{\rho \beta_j \beta}{\bar{\gamma}_{SR}\bar{\gamma}_{RD}}}\right), \tag{20}$$

where $S_{\mu,\nu}(\cdot)$ represents Lommel function defined in [12, (8.57)]. A closed-form solution for I_B from (19) is also found in [11]:

$$I + B = \frac{-1}{2\ln(2)} \sum_{i=0}^{k-1} \frac{\alpha_i}{\beta_i} \exp\left(\frac{\beta_i \rho}{\bar{\gamma}_{RD}}\right) E_1\left(\frac{\beta_i \rho}{\bar{\gamma}_{RD}}\right). \tag{21}$$

The term I_y in (19) could not be derived as a closed-form solution, but it can be upper bounded using again Jensen's inequality, thus obtaining [11]:

$$I_y \leq \frac{1}{\bar{\gamma}_{SR}\ln(2)} \sum_{i=0}^{k-1} \alpha_i \exp\left(\frac{\beta_i}{\bar{\gamma}_{SR}}\right) \sum_{j=0}^{k-1} \frac{\alpha_j}{\beta_j} \exp\left(\frac{\rho \beta_j}{\bar{\gamma}_{RD} L(i)}\right) \Gamma\left(0, \frac{\rho \beta_j}{\bar{\gamma}_{RD} L(i)}\right), \tag{22}$$

where $\Gamma(\cdot, \cdot)$ denotes the upper incomplete Gamma function defined in [10, (8.350.2)]. $L(i)$ is equal to:

$$L(i) = \left(\frac{\bar{\gamma}_{SR}}{\beta_i}\right)^2 - \left[\frac{\bar{\gamma}_{SR}}{\beta_i} + \left(\frac{\bar{\gamma}_{SR}}{\beta_i}\right)^2\right]^2 \exp\left(-\frac{\beta_i}{\bar{\gamma}_{SR}}\right). \tag{23}$$

The obtained upper bound for the I_y actually presents the term which defines the lower bound for the ergodic capacity of the k-th subcarrier in the considered OFDM AF FG relay system with BTB SCP, and in final, through (18) it gives the lower bound of the ergodic capacity of the whole considered system.

4 Results

The subsequent presented analytical and simulation results assume perfectly synchronized dual-hop OFDM AF FG relay system implementing SCM. It is assumed that the noise variances at R and D are equal, i.e. $N_{02} = N_{02}$, as well as the average symbol energies transmitted by S and by R, $\epsilon_S = \epsilon_R$. The so-called semi-blind scenario is considered, where R uses knowledge on channel state information (CSI) about the S-R link to calculate the gain G:

$$G^2 = \mathbf{E}\left[\frac{\epsilon_R}{|H_{1,k}|^2 \epsilon_S + N_{01}}\right], \tag{24}$$

yielding

$$G^2 = \frac{\epsilon_R}{\epsilon_S \mathbf{E}[|H_{1,k}|^2]} e^{1/\bar{\gamma}_{SR}} E_1\left(\frac{1}{\bar{\gamma}_{SR}}\right). \tag{25}$$

Simulation results are obtained through Monte Carlo simulations of the part of the OFDM system that belongs to frequency domain, what is possible and accurate approach as we have assumed perfect synchronization among the communication terminals. The subcarrier channel transfer functions on both hops are generated as independent Gaussian complex random variables with zero mean.

4.1 BER Performance

BER performances of DPSK modulated OFDM AF FG relay systems with SCM, as a function of total transmitted power per subcarrier P_T, are given in Figure 2. In order to get an insight in power balance of OFDM AF FG relay system with SCM, comparison of BER performances, with the OFDM AF FG relay system and with the direct transmission scenario are also provided. We have chosen parameters $A_1 = 2$ and $A_2 = 10$, which for downlink communication corresponds to the scenario where the distance between the R and D terminals is shorter than the one between the S and R terminals. For a direct transmission scenario A_{eq} is calculated using (13).

From the given BER plots, the advantage of using relay transmission over the direct transmission is clear, as the power saving per subcarrier is about 2 dB for the BER values lower than 10^{-2}. Comparing to basic OFDM AF FG relay system, additional 1dB of transmitted power can be saved for achieving the same BER performances if BTW SCM scheme is used, for all the BER values below 10^{-1}. When BTB SCM scheme is concerned, it is shown that it achieves the best BER performances in the regime of very low transmitted power, i.e. in the cases where the average SNRs on both hops are low. In the assumed scenario, for P_T values above 15 dB, the system with BTB SCM scheme has worse BER performances even than the system with direct transmission. Thus, in order to achieve the best BER performance, for

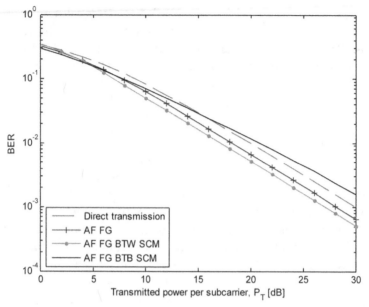

Figure 2 BER for DPSK modulated OFDM system and OFDM AF FG relay system with and without SCM, as a function of P_T.

the same transmitted energy, in the region of small SNRs BTB SCM scheme should be implemented, while for the average and medium values of SNRs, BTW SCM scheme should be applied.

Figure 3 gives analytical and simulation results for DPSK modulated OFDM AF FG relay system implementing both BTB SCM and BTW SCM, in a scenario with equal average SNR values on both hops. For the sake of comparison, BER performance of the ordinary OFDM AF FG relay system is also presented. As it is obvious that in the relay system implementing BTB SCM, the subcarriers with the lowest SNRs on both hops, which are mapped at R station creating subcarrier pair $k = 1$, have the greatest contribution on increase of BER values, than we present the BER performances of relay system employing BTB SCM without "worst" subcarrier pair (w/o $k = 1$), for a different total of numbers of subcarriers in the system ($M = 16, 32, 64$).

The presented results confirm the accuracy of the undertaken analytical approach, as we have excellent matching between the analytical results and simulation results for all SNR values. As expected, in the region of low values of average SNR per hop, BTB SCM scheme achieves the best BER performance. It outperforms the system with BTW SCM scheme for the values

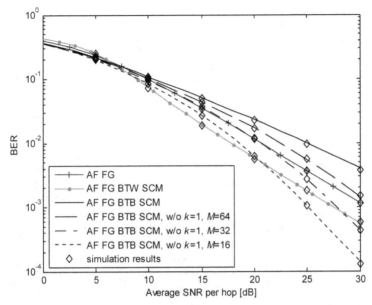

Figure 3 BER for DPSK modulated OFDM AF FG relay system with and without SCM.

of average SNR per hop up to approximately 6.5 dB. For higher values of average SNR system with BTW SCM has the lowest BER, and its advantage in BER performances increases very fast as the values of average SNRs increase. Thus for example, for the BER value of 10^{-2}, it already achieves SNR gain of 3 dB over the OFDM AF FG relay system and more than 7 dB SNR gain in comparison with the system implementing BTB SCM.

    Let us now consider the system implementing BTB SCM, where the worst subcarrier pair is not used, From Figure 3 it can be seen that, besides for the very small values of average SNR per hop, only in the region of very high values of average SNR per hop (about 29 dB for $M = 32$ and 20 dB for $M = 16$), and for the cases of smaller number of subcarriers in OFDM system, this solution outperforms the system with BTW SCM in terms of BER. However, having in mind that in the real-case scenarios, such a big values of average SNRs on both hops will appear with very small probability, and that the OFDM systems with $M = 16$ and $M = 32$ subcarriers are not of the great practical importance, we wanted to examine if omitting several "worst" subcarrier pairs in the OFDM systems with greater number of subcarriers may bring enough BER performance improvement to have one SCM as an optimal solution for all SNR values of interest.

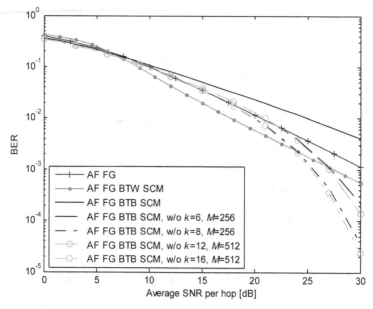

Figure 4 BER for DPSK modulated OFDM AF FG relay system with and without SCM.

Figure 4 presents BER performance of OFDM AF FG relay system with SCM having $M = 256$ and $M = 512$ subcarriers, where in the case of BTB SCM having $M = 256$ subcarriers $k = 6$ and $k = 8$ worst subcarrier pairs are not used, while in the system with $M = 512$ subcarriers, $k = 12$ and $k = 16$ worst subcarrier pairs are omitted. The given results confirm again that for the SNR values of interest, for the BER performance improvement in OFDM AF FG relay systems, BTB SCM should be implemented in the regions of small values of average SNRs on both hops, and BTW SCM should be implemented in the region of medium and high values of average SNRs on hops.

4.2 Ergodic Capacity

Figure 5 presents the obtained lower and upper bounds of the ergodic capacity for the considered system, as well as the ergodic capacity graph obtained through simulations. In order to identify the level of capacity improvement achieved through implementation of BTB SCM scheme, the simulation results for the OFDM AF FG relay system without (w/o) SCM are also given. The average ergodic capacity graphs are presented as a function of SNR

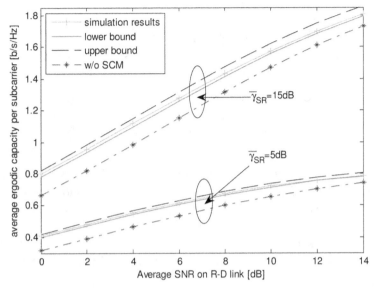

Figure 5 Average ergodic capacity of OFDM AF FG relay system with and without BTB SCM.

on R-D link, for two different SNR values on S-R link, $\bar{\gamma}_{SR} = 5$ dB and $\bar{\gamma}_{SR} = 15$ dB.

The results presented in Figure 5 approve that the obtained analytical results present very tight upper and lower bounds, as they deviate very little from the simulation obtained results. This is more obvious through the data given in Table 1.

Table 1 presents values of the average ergodic capacity of OFDM AF FG relay system with BTB SCM, obtained through simulations and through analytical calculations of lower and upper bounds, as well as the ratios of analytically obtained results to simulation results expressed in percentages. For the sake of comparison, average ergodic capacity values of the OFDM AF FG relay system without SCM are also included, and the percentages of ergodic capacity enhancement attained through BTB SCM.

First of all, Table 1 shows that implementing BTB SCM can enhance ergodic capacity of OFDM AF FG relay system up to 30%, and the highest capacity enhancement is achieved in the region of small average SNR values on both hops. This is very important as it means the SCM technique may achieve the requested QoS in the case of very bad channel conditions on both hops, i.e. in the scenario where the regular OFDM AF FG relay system

Table 1 Average ergodic capacity values for $\bar{\gamma}_{SR} = 5$ dB.

$\bar{\gamma}_{RD}$ [dB]	0 dB	6 dB	14 dB
Simul. res. [b/s/Hz]	0.406	0.618	0.786
w/o SCP [b/s/Hz]	0.318	0.535	0.744
% cap. enhancement	**27.7%**	**15.5%**	**5.6%**
Low. bound [b/s/Hz]	0.397	0.608	0.779
% of simul. res.	97.78%	98.38%	99.1%
Upp. bound [b/s/Hz]	0.417	0.633	0.805
% of simul. res.	102.7%	102.4%	102.4%

may fail in fulfilling end-user's demands. Besides this, from Table 1 we can see that the derived upper bound deviate from the simulation results up to 2.7%, while the derived lower bound differs from the simulation results up to 2.12%. Both these results confirm the accuracy of the undertaken analytical approaches, and the obtained ergodic capacity bounds may be effectively used for the assessment of the performance of OFDM AF FG relay system with SCM.

5 Conclusions

We analyzed the level of BER performance improvement and ergodic capacity enhancement achieved through implementation of ordered subcarrier mapping (SCM) at R station of OFDM AF FG relay system, which represents a very interesting solution for the upcoming specifications of IMT-Advanced mobile cellular systems. It has been shown analytically, and completely verified through simulations, that significant BER performance improvement may be achieved in the considered relay system if BTB SCM is implemented in the region of small values of average SNRs on both hops, and BTW SCM scheme in the region of medium and high values of average SNRs on both hops. Furthermore, we have shown that in OFDM AF FG relay system with BTB SCM, even when several subcarriers with the lowest SNRs from the both hops are not used, the BER performance in the region of medium and high values of average SNRs, which are of interest, cannot be improved enough to outperform the system implementing BTW SCM. Additionally, we presented energy efficiency of the considered SCM solution through presenting BER performance improvement achieved for the same power transmitted per subcarrier, in comparison with the ordinary OFDM AF FG relay system and with the direct case transmission.

When ergodic capacity of the system is considered, we analyzed only the solution assuming implementation of BTB SCM, which is the SCM scheme that is proven to maximize capacity for all the SNR values on both hops. We derived closed-form analytical expressions representing tight upper and lower bounds of the achievable ergodic capacity of the OFDM AF FG relay system with BTB SCM. These bounds can be efficiently used for the assessment of the ergodic capacity of the considered system, as they deviate from the simulation results less than 3% for all the SNR values. Ergodic capacity performance comparison of the analyzed relay system implementing BTB SCM, with the relay system without SCM has shown that ergodic capacity may be enhanced up to 30% in the region of small values of average SNRs on both hops. This is particularly important as it means significant capacity enhancement in the scenario with bad channel conditions on both hops.

All the presented BER and ergodic capacity performance have approved that dual-hop OFDM AF FG relay system with SCM at R station, represents very interesting energy efficient solution for the next generation mobile cellular systems.

References

[1] E. Yang, H. Hu, J. Xu, and G. Mao. Relay technologies for WiMAX and LTE-advanced mobile systems. IEEE Commun. Magazine, 47(10): 100–105, October 2009.

[2] A. Hottinen and T. Heikkinen. Subchannel assignment in OFDM relay nodes. In Proceedings of the 40th Annual Conference on Information Sciences and Systems, pp. 1314–1317, 2006.

[3] I. Hammerstrom and A. Wittneb. Joint power allocation for non-regenerative MIMO-OFDM relay links. In Proceedings of the IEEE International Conference on Acoustic, Speech and Signal Processing, Vol. 4, p. IV, May 2006.

[4] M. Herdin. A chunk based OFDM amplify-and-forward relaying scheme for 4G mobile radio systems. In Proceedings of the IEEE ICC 2006, Istanbul, Turkey, 2006.

[5] C.K. Ho and A. Pandharipande. BER minimization in relay-assisted OFDM systems by subcarrier permutation. In Proceedings of the IEEE VTC08, pp. 1489–1493, Singapore, 2008.

[6] E. Kocan, M. Pejanovic-Djurisic, D.S. Michalopoulos, and G.K. Karagiannidis. BER performance of OFDM amplify-and-forward relaying system with subcarrier permutation. In Proceedings of IEEE Wireless VITAE 2009 Conference, Aalborg, Denmark, pp. 252–256, May 2009.

[7] E. Kocan, M. Pejanovic-Djurisic, D.S. Michalopoulos, and G.K. Karagiannidis. Performance evaluation of OFDM amplify-and-forward relay system with subcarrier permutation. IEICE Trans. on Commun., E93-B(05):1216–1223, May 2010.

[8] E. Kocan, M. Pejanovic-Djurisic, and G.K. Karagiannidis. New solution for BER per-
     formance improvement of OFDM AF relay systems. In Proceedings of IEEE Conference
     TELFOR, Belgrade, Serbia, pp. 412–415, October 2012.
[9] M.K. Simon and M.-S. Alouini. Digital Communication over Fading Channels, 2nd ed.
     Wiley, New York, 2005.
[10] M. Abramovitz and I.A. Stegun. Handbook of Mathematical Functions with Formulas,
     Graphs, and Mathematical Tables, 9th ed. Dover, New York, 1972.
[11] E. Kocan, M. Pejanovic-Djurisic, and Z. Veljovic. Ergodic capacity of OFDM AF fixed
     gain relay system with subcarrier mapping. In Proceedings of IEEE Conference WTS
     2012, London, April 2012.
[12] I.S. Gradshteyn and I.M. Ryzhik. Table of Integrals, Series, and Products, 6th ed.
     Academic Press, New York, 2000.

Biographies

Milica Pejanovic-Djurisic is Full Professor in Telecommunications at the University of Montenegro, Faculty of Electrical Engineering, Podgorica, Montenegro. Professor Pejanovic-Djurisic graduated in 1982 from University of Montenegro with BSc degree in Electrical Engineering. She received her MSc and PhD degrees in Telecommunications from University of Belgrade. For a period of two years, Professor Pejanovic-Djurisic also performed research in mobile communications at University of Birmingham, UK. She has been teaching at University of Montenegro telecommunications courses on graduate and postgraduate levels, being the author of four books, many strategic studies, and participating in numerous internationally funded research teams and projects. She has published more than 200 scientific papers in international and domestic journals and conference proceedings. Professor Pejanovic-Djurisic has organized several workshops and given tutorials and speeches at many scientific and technical conferences. Her main research interests are: wireless communications theory, wireless networks performance improvement, broadband transmission techniques, optimization of telecommunication development policy. She has considerable industry and operating experiences working as industry consultant and Telecom Montenegro Chairman of the Board. Professor Pejanovic-Djurisic has also been involved in activities related with telecommunication regulation. Being an ITU expert, she participates in a number of missions and ITU workshops related with regulation issues, development strategies and technical solutions.

Enis Kocan is a teaching/research assistant at the University of Montenegro, Faculty of Electrical Engineering, Podgorica, Montenegro. He received the

BSc and MSc degrees in electronics engineering from the University of Montenegro, in 2003 and 2005, respectively. He defended his Ph.D. thesis at the same University in 2011, in the area of mobile communications, with the topic being OFDM-based cooperative communications for future generation mobile cellular systems. His major research interests are in digital communications over fading channels, physical layer aspects of wideband cooperative systems and multi-hop communications. Dr. Kocan has published and presented more than 40 scientific papers in international and national scientific journals, international and regional conferences and is co-author of the book *OFDM-Based Relay Systems for Future Wireless Communications*, published by River Publishers, Denmark in 2012.

Improving Energy Efficiency of Relay Systems Using Dual-Polarized Antenna

M. Ilic-Delibasic and M. Pejanovic-Djurisic

Faculty of Electrical Engineering, University of Montenegro, Podgorica, Montenegro; e-mail: {majai, milica}@ac.me

Received 15 November 2012; Accepted 20 December 2012

Abstract

In this paper we study the bit error analyses of decode-and-forward (DF) relay system over Ricean fading channels, implementing polarization diversity at the relay node. Two correlated and non-identical Ricean fading channels are used to describe source to relay link, and maximal ratio combining (MRC) of the received signals is performed at the relay. Communication between the relay and destination is also assumed to be over Ricean fading channel. The performances of the considered system are compared to classic DF relay system with single antenna relay, in order to determine the level of BER (Bit Error Rate) improvement. A significant decrease of the average SNR per hop in order to achieve the same BER performance is realized using dual diversity at relay, despite certain levels of correlation and power unbalance between the diversity branches, which can further reduce the needed transmit power.

Keywords: Energy efficiency, decode-and-forward, relay, polarization diversity, bit error rate.

1 Introduction

In the past few decades wireless communications industry witnessed tremendous growth. Many new standards and network topologies have been developed in order to support more demanding applications. Along with

Journal of Green Engineering, Vol. 3, 167–179.

providing service to great number of subscribers with desirable throughput, service providers are also facing with significant energy consumption. The rising energy costs and carbon footprint of wireless networks continuously motivate research community, network operators and regulatory bodies to address energy-efficiency as an important issue in exploring new solutions for future wireless systems.

Recently, relay based communication systems have received a significant attention by both industry and academia due to their ability to improve wireless link performance and increase capacity of a wireless system. Furthermore, by using intermediate relays (R) to assist transmission from the source (S) to the destination (D), such systems can increase network coverage and enable more energy efficient service. They can also offer a virtual antenna array bringing spatial diversity benefits while still satisfying the size and power constraints of mobile devices [1–4]. In fact, one of the main advantages of such communication technique is that it distributes the use of power throughout the hops, reducing the need to use a large power at the transmitter. This can provide a longer battery life of mobile terminals and lower level of interference introduced to the rest of the network [3]. In order to enable further energy savings, different energy aware schemes can be used, such as different cooperative algorithms, power allocation, relay selection, sharing or distributing tasks among cooperating entities, etc.

Multiple antennas (diversity) are taken as another enabling technology to increase the wireless system performance. Recently, multi-antenna relaying systems have been proposed [5–8] implementing multiple antennas at S, R and/or D. However, the deployment of multiple antennas at wireless terminals may not be feasible in some scenarios, especially when future terminals are expected to be small and light. That is why multiple antennas are usually implemented at the relays (infrastructure based fixed relay) instead of terminals [5–7]. However, due to a limited space, it is still not always easy to accommodate multiple antennas. In this manner collocated antennas could be interesting for both relay and mobile terminals, but this makes diversity system worthless. However, using dual polarized antenna can be rather practical, because two orthogonally polarized antennas can be collocated. Different propagation characteristics in wireless communication systems for vertically and horizontally polarized waves enable implementation of single antenna structure employing two orthogonal polarizations. Multiple reflections of electromagnetic wave from source to destination depolarize and decorrelate signals, while different propagation characteristics for vertically (V) and horizontally (H) polarized waves make diversity possible.

This is why in this paper we assume that the source and the destination terminals are equipped with a single antenna, while relay terminal has dual-polarized receiving and single transmitting antenna. We use maximal ratio combining (MRC) of the received signals at the R, and regenerative decode-and-forward (DF) relay system which fully decodes the data packet at the relay prior to transmission over the second hop. Non-identical Ricean fading environment is assumed, since in many practical scenarios source-relay (S-R) link can experience different fading conditions than relay-destination (R-D) link. A lot of practical channel measurement results [9] have validated such asymmetric nature of the relay fading channels, termed as mixed type fading channels [10]. In contrast to the symmetric fading assumptions [11–13], there are few performance analysis results for mixed type fading channels [10, 14]. Despite the importance of the Ricean model, only a few works have analyzed the performance of relays under LoS (Line of Sight) fading conditions.

The paper is organized as follows. Following the introduction, a detailed description of the system model is given in Section 2, followed by the mathematical analysis in Section 3. The obtained results are presented in Section 4, and concluding remarks are given in Section 5.

2 System Model

In this paper we considered dual hop relay system where source communicate to destination via relay. R is assumed to be equipped with one transmit and dual-polarized receive antenna. Due to the half-duplex nature of operation of the relay, transmission protocol of the relaying paths divides each transmission period into two time intervals: in the first time slot S communicates with R. Relay then receives the signals from S through two orthogonally polarized links and performs MRC of the received signals, decodes and then forwards to D. Analyzed system is illustrated in Figure 1.

2.1 Polarization Diversity

It is well known that the usage of multiple antennas on one or both ends of a wireless link enables necessary improvements of the system performance. The most commonly used diversity technique is space diversity. However, for reaching a full diversity gain, the received antennas must be adequately separated. At the same time, a space for mounting two or more antennas can be limited. A promising alternative for overcoming such a problem is polarization diversity [15]. Since it utilizes only one dual polarized antenna,

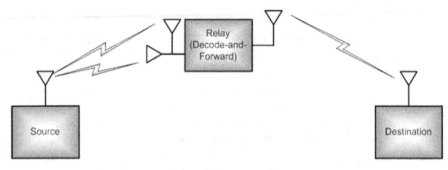

Figure 1   System model.

this type of diversity is space and cost effective solution despite the fact that the collocation of antennas necessary leads to signal correlation.

Polarization diversity is based on the fact that propagation characteristics in wireless communication system are different for vertically and horizontally polarized waves. Multiple reflections between the transmitter and the receiver lead to the depolarization of radio waves, coupling some energy into the orthogonal polarization. Due to that characteristic of multipath radio channel, vertically/horizontally polarized transmitted waves have also horizontally/vertically component.

There are two main parameters describing polarization diversity: cross-polar discrimination (XPD), that indicates power difference between the average powers of copolarized and crosspolarized signals, and correlation coefficient between the received signal envelopes. Many measurements conducted in different environments show that typical values of XPD vary from 1-10 dB in urban and suburban environment, and 10–18 dB in rural environment [16]. Correlation coefficient is usually very small (generally even less than 0.3), and does not significantly affect system performances.

2.2 Channel Model

In this paper we consider Ricean fading channel, which is often used to model propagation paths consisting of one strong direct line-of-sight signal and many randomly reflected and usually weaker signals. It is a very useful and widely used channel model as it spans in range from Rayleigh fading to no fading cases [17].

In many practical scenarios S-R and R-D links can be characterized with asymmetric fading conditions, for example, one link may have a strong line-of-sight component (typical Ricean fading), while the other link may have greater severity of fading (closer to Rayleigh) [10]. We assume that there is no direct communication between source and destination, and R is equipped with a dual polarized antenna, with MRC of the received signals.

In order to analytically describe such system, taking into account the existing signal correlation between two diversity branches available at R, first in [18] we propose an analytical method for transforming the correlated signals into uncorrelated ones. In that manner, effects of the branch correlation are expressed through modification of the following system parameters: the average received signal-to-noise ratio (SNR) per diversity branch and Ricean K factor. According to the procedure completely described in [18], the transformed average SNRs and K factors are given by

$$\bar{\gamma}_{iT} = a_i^2 \bar{\gamma}_1 + (1 - a_i^2)\bar{\gamma}_2 + 2a_i \sqrt{\frac{(1 - a_i^2)\bar{\gamma}_1\bar{\gamma}_2}{(1 + K_1)(1 + K_2)}} (\rho + \sqrt{K_1 K_2}) \quad (1)$$

and

$$K_{iT} = \frac{\frac{a_i^2 X K_1}{1+K_1} + \frac{(1-a_i^2)K_2}{1+K_2} + 2a\sqrt{\frac{(1-a_i^2)XK_1K_2}{(1+K_1)(1+K_2)}}}{\frac{a_i^2 X}{1+K_1} + \frac{1-a_i^2}{1+K^2} + 2\rho a_i\sqrt{\frac{(1-a_i^2)X}{(1+K_1)(1+K_2)}}} \quad (2)$$

with a_i [18]:

$$a_i = \sqrt{\frac{1}{2} - \frac{(-1)^i}{2}\sqrt{\frac{\left(\frac{X}{1+K_1} - \frac{1}{1+K_2}\right)^2}{\left(\frac{X}{1+K_1} - \frac{1}{1+K_2}\right)^2 + \frac{4\rho^2 X}{(1+K_1)(1+K_2)}}}} \quad (3)$$

K_i and $\bar{\gamma}_i$ ($i = 1, 2$) are the original channel parameters: Ricean factors and average SNR per diversity branch, ρ is correlation coefficient between the signals, and X denotes XPD, that is the level of power imbalance, i.e. $X = \bar{\gamma}_1/\bar{\gamma}_2$.

After the proposed transformation, standard mathematical model for performance analysis of uncorrelated signals can be applied.

2.3 DF Relay System

Implementation of relays in wireless network extends coverage of a base station (BS). Considering well-known properties of the wireless channel, including large path losses, shadowing and multipath fading, covering very distant users via direct transmission becomes very demanding in terms of required power for establish a reliable connection. High power transmission requirement further claims high power consumption and also introduces high level of interference. Dividing a direct path between S and D into several shorter links (hops) using relays reduces signal attenuation, and thus reduces the transmitting power necessary to achieve a same signal-to-noise ratio (SNR) level. This is the key point which makes relay systems a good candidate for green communications.

In this paper a decode-and-forward (DF) technique is employed in dual-hop relay communication systems. DF has several advantages over another well-known relay technique: the amplify-and-forward (AF). One of these advantages is the fact that R first decodes the received signal, re-encodes it and then transmits to D. In this manner, the total noise at D is decreased compared to AF which amplifies the noise along with the signal. Another advantage of such system is that it is possible to implement different modulation schemes at S-R and R-D links, depending on the link conditions. However, DF signal processing at R can have certain drawbacks, especially in the case of channels with severe fading. When BER is concerned, if the decoding error occurs at R, those erroneous symbols will be further transmitted to D. There is no doubt that the characteristics of the R-D link contribute to an overall BER performance degradation of DF relay system. In addition, errors might be also introduced in the signal at terminal D.

In order to reduce those negative implications of the error propagation and to improve BER performance, several techniques can be applied. Different encoding schemes for error detection and correction can be applied in the process of signal regeneration at DF relay stations [19]. Knowing that diversity systems are a standard solution for BER performance improvement, we propose to implement diversity combining at R.

3 Performance Analysis

In DF relay systems a signal is transmitted over two cascade links and its decoding is done twice. If the transmission implies a binary signal with two possible symbol states (DPSK or BPSK), an error will appear at the final destination terminal only if an error in the signal detection is performed once (either on the first or on the second link):

$$P = P_{SR}(1 - P_{RD}) + (1 - P_{SR})P_{RD}, \tag{4}$$

where P_{SR} and P_{RD} are the bit error rates at the S-R and R-D links, respectively. In order to determine PSR and PRD, we use the MGF (Moment Generating Function) approach [17].

If the MRC of the received diversity signals at R is applied, the MGF of total received SNR at R can be obtained as a product of individual MGFs, in the case that the received signals are independent. If we apply the above described transformation of correlated signals into uncorrelated ones, proposed in [18], the MGF for S-R link in the case of Ricean fading channels becomes

$$\mathrm{MGF}_{SR}(s) = \frac{1 + K_{1T}}{1 + K_{1T} - s\bar{\gamma}_{1T}} \frac{1 + K_{2T}}{1 + K_{2T} - s\bar{\gamma}_{2T}}$$

$$\times \exp\left(\frac{K_{1T}s\bar{\gamma}_{1T}}{1 + K_{1T} - s\bar{\gamma}_{1T}} + \frac{K_{2T}s\bar{\gamma}_{2T}}{1 + K_{2T} - s\bar{\gamma}_{2T}}\right), \qquad (5)$$

while the MGF for the S-R link, also in the case of Ricean fading is given as [17]

$$\mathrm{MGF}_{RD}(s) = \frac{1 + K_{RD}}{1 + K_{RD} - s\bar{\gamma}_{RD}} \exp\left(\frac{K_{RD}s\bar{\gamma}_{RD}}{1 + K_{RD} - s\bar{\gamma}_{RD}}\right) \qquad (6)$$

Knowing the MGF, the probability of error in the case of BPSK mapping can be determined as [17]

$$P_{\mathrm{BPSK}} = \frac{1}{\pi} \int_0^{\pi/2} \mathrm{MGF}\left(-\frac{1}{\sin^2 t}\right) dt \qquad (7)$$

Putting (5) and (6) into (7) P_{SR} and P_{RD} are determined. Finally, the overall BER is determined using (4).

In the case of DPSK mappings, BER can be determined as:

$$P = 0.5\mathrm{MGF}(1). \qquad (8)$$

Using (5), (6) and (4), BER can be determined as

$$P_{\mathrm{DPSK}} = 0.5[\mathrm{MGF}_{SR}(1) - (1 - 0.5\mathrm{MGF}_{RD}(1)) + (1 - 0.5\mathrm{MGF}_{SR}(1_{\mathrm{MGF}_{RD}}(1)].$$
$$(9)$$

Or, equivalently:

$$P_{\mathrm{DPSK}} = 0.5[\mathrm{MGF}_{SR}(1) + \mathrm{MGF}_{RD}(1) - \mathrm{MGF}_{SR}(1)\mathrm{MGF}_{RD}(1)]. \qquad (10)$$

4 Results

When system with polarization diversity is analyzed, it has been shown that it is necessary to take into account the influence of the two main parameters characterizing this type of diversity: correlation coefficient between the received signals and XPD. These two parameters are strongly dependent on the environment, thus greatly affect system performance. However, various experimental results have shown that typical values for correlation coefficient vary from 0.2–0.4 [16], while studies have shown that systems based on multiple antennas can achieve a significant performance improvement as long as the correlation coefficient is less than 0.7 [20]. The other very important parameter in assessing the effectiveness of the polarization diversity system is XPD. This parameter is strongly dependent on environment, and has larger influence on overall system performance than correlation coefficient. High XPD increases BER, and if too large can make the whole diversity system worthless. As it is already noted, typical values of this parameter vary from 1 to 10 dB in urban and suburban environments, and from 10 to 18 dB in rural environments [16].

Still, it has been shown that despite a certain level of the correlation between the signals from different diversity branches and power imbalance between them, polarization diversity system provides significant performance improvements of wireless systems. It can be considered as an alternative to commonly used space diversity, especially in urban environments with the lack of space to mount two or more spatially separated antennas. Fortunately, in urban environments the effect of depolarization is stronger, leading to lower XPD, thus better performances. Since it utilizes only one dual polarized antenna, this type of diversity presents an attractive space and cost effective solution [18]. That motivated us to analyze achievable level of performance improvement using polarization diversity at R.

Some of the BER results obtained using the analytical model described in the previous section, are illustrated in Figure 2 for BPSK and in Figure 3 for DPSK mapping. In order to determine the level of BER improvement obtained by implementing dual polarized receive antenna at R, BER curve for the relay system with no diversity at R is also presented. It is assumed that the average total received SNR at the output of the MRC combiner has the same value as in the case with no diversity. Equal average SNRs per hop is assumed. In order to verify the proposed analytical model for BER determination of the analyzed wireless relay system with polarization diversity implemented at R, the above presented mathematical results are compared with the results obtained with the adequate simulation model of the analyzed system.

It has been showed that implementing dual-polarized antenna at the R leads to significant BER improvements. As it could have been expected, the level of the achieved performance enhancements depends on power differ-

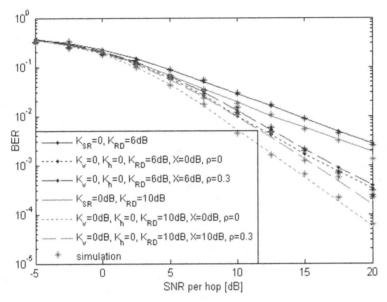

Figure 2 BER performances of DF relay system, BPSK.

ence between diversity branches and correlation coefficient. However, still, for some typical XPD and ρ values, the improvement cannot be neglected.

We also considered the case when average SNR at S-R link is not necessary equal to the average SNR at R-D link. Three scenarios are presented: the average SNR of S-R link is greater, equal or less than the average SNR of R-D link. In all the considered scenarios, performance improvement is evident, as shown in Figure 4.

From the results presented it can be concluded that in all considered cases, implementing diversity at R can significantly contribute to energy savings. For example, in order to achieve BER of the order of 10^{-3} up to 5 dB lower SNR is required, implying a lower transmit power.

5 Conclusions

Motivated by the importance of relaying systems, we analyzed in this paper the possibility for BER performance improvement of DF relay system through implementation of diversity reception on relay. Polarization diversity provides the possibility to implement two collocated antennas with sufficiently low correlation coefficient. However, due to the existing correlation,

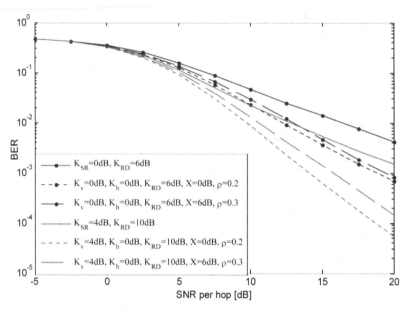

Figure 3 BER performances of DF relay system, DPSK.

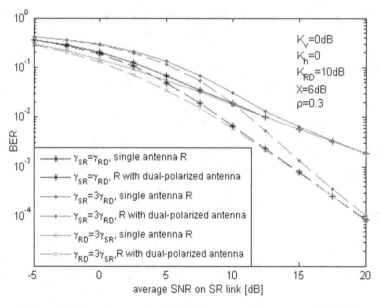

Figure 4 BER performances of DF relay system ($\bar{\gamma}_{SR}$ constant), BPSK.

and SNR unbalance per diversity branch, the analytical model describing such system is very complex. This is why we used a previously defined linear transformation of correlated signals into uncorrelated ones, and applied a standard MGF approach for conditional BER.

The obtained results show significant performance improvements for various XPD and ρ. In addition, such a system further improves energy efficiency of classical relay systems. Namely, splitting the communication link into several shorter hops, instead of direct communication between the S and D, the transmitted power is effectively distributed throughout the system. Further enhancement means that lower transmitted power is needed for equal system performance, with only slight modification at the relay. This implies a more energy efficient system that will provide longer battery lifes for mobile terminals, which is considered as a first step towards green communications.

References

[1] E.C. van der Meulen. Three-terminal communication channels. Advanced Applied Probability, 3: 120–154, 1971.

[2] A. Nosratinia and A. Hedayat. Cooperative communications in wireless networks. IEEE Communications Magazine, 74–80, October 2004.

[3] J.N. Laneman and G.W. Wornell. Energy-efficient antenna sharing and relaying for wireless networks. In Proceedings of IEEE WCNC 2000, Vol. 1, pp. 7–12, October 2000.

[4] R. Pabst, B.H. Walke, D.C. Schultz, P. Herhold, H. Yanikomeroglu, S. Mukherjee, H. Viswanathan, M. Lott, W. Zirwas, M. Dohler, H. Aghvami, D.D. Falconer, and G.P. Fettweis. Relay-based deployment concepts for wireless and mobile broadband radio. IEEE Communications Magazine, 42(9): 80–89, September 2004.

[5] A. Adinoyi and H. Yanikomeroglu. Cooperative relaying in multiantenna fixed relay networks. IEEE Trans. Wireless Commun., 6(2): 533–544, February 2007.

[6] H. Katiyar and R. Bhattacharjee. Performance of two-hop regenerative relay network under correlated Nakagami-m fading at multi-antenna relay. IEEE Communications Letters, 13(11): 820–822, November 2009.

[7] H. Katiyar and R. Bhattacharjee. Performance of MRC combining multi-antenna cooperative relay network. International Journal of Electronics and Communications (AEU), 64: 988–991, 2010.

[8] S. Prakash and I. McLoughlin. Performance of dual-hop multi-antenna systems with fixed gain amplify-and-forward relay selection. IEEE Transactions on Wireless Communications, 10: 1709–1714, June 2011.

[9] Multi-hop relay system evaluation methodology (channel model and performance metric). Technical Report, IEEE 80216j-06-013r3, February 2007.

[10] H.A. Suraweera, R.H.Y. Louie, Y. Li, G.K. Karagiannidis, and B. Vucetic. Two hop amplify-and-forward transmission in mixed Rayleigh and Rician fading channels. IEEE Commun. Lett., 13: 227–229, 2009.

[11] M.O. Hasna and M.S. Alouini. End-to-end performance of transmission systems with relays over Rayleigh-fading channels. IEEE Trans. Wirel. Commun., 2: 1126–1131, 2003

[12] T.A. Tsiftsis, G.K. Karagiannidis, P.T. Mathiopoulos, and S.A. Kotsopoulos. Nonregenerative dual-hop cooperative links with selection diversity. EURASIP J. Wireless Commun. Netw., 2006: 34–34, 2006.

[13] M.O. Hasna and M.S. Alouini. Harmonic mean and end-to-end performance of transmission systems with relays. IEEE Trans. Commun., 52: 130–135, 2004.

[14] W. Xu, J. Zhang, and P. Zhang. Performance analysis of dual-hop amplify-and-forward relay system in mixed Nakagami-m and Rician fading channels. Electronics Letters, 46(17), August 2010.

[15] W.C.Y. Lee and Y.S. Yeh. Polarization diversity system for mobile radio. IEEE Transactions on Communications, COM-20(5), October 1972.

[16] J. Jootar and J.R. Zeidler. Performance analysis of polarization receive diversity in correlated Rayleigh fading channels. In Proceedings of IEEE Globecom Conference, pp. 774–778, November 2003.

[17] M.K. Simon and M.S. Alouini. Digital Communication over Fading Channels, 2nd ed. Wiley-Interscience, 2005.

[18] M. Ilic-Delibasic, M. Pejanovic-Djurisic, and R. Prasad. A novel method for performance analysis of ofdm polarization diversity system in Ricean fading environment. Wireless Personal Communications, 63(3): 751–764, April 2012.

[19] T. Wang, A. Cano, G.B. Giannakis, and J.N. Laneman. High-performance cooperative demodulation with decode-and-forward relays. IEEE Transactions on Communications, 55(7), July 2007.

[20] L.C. Lukama and D.J. Edvards. Performance of spatial and polarization diversity. In Proceedings of Wireless Personal Multimedia Communications (WPMC01), Aalborg, Denmark, 2001.

Biographies

Milica Pejanovic-Djurisic is Full Professor in Telecommunications at the University of Montenegro, Faculty of Electrical Engineering, Podgorica, Montenegro. Professor Pejanovic-Djurisic graduated in 1982 from University of Montenegro with BSc degree in Electrical Engineering. She received her MSc and PhD degrees in Telecommunications from University of Belgrade. For a period of two years, Professor Pejanovic-Djurisic also performed research in mobile communications at University of Birmingham, UK. She has been teaching at University of Montenegro telecommunications courses on graduate and postgraduate levels, being the author of four books, many strategic studies, and participating in numerous internationally funded research teams and projects. She has published more than 200 scientific papers in international and domestic journals and conference proceedings. Professor Pejanovic-Djurisic has organized several workshops and given tutorials and speeches at many scientific and technical conferences. Her main research interests are: wireless communications theory, wireless networks performance improvement, broadband transmission techniques, optimization of telecommunication development policy. She has considerable industry and operating experiences working as industry consultant and Telecom Montenegro Chairman of the Board. Professor Pejanovic-Djurisic has also been involved in activities related with telecommunication regulation. Being an ITU expert, she participates in a number of missions and ITU workshops related with regulation issues, development strategies and technical solutions.

Maja Ilic-Delibasic is a teaching/research assistant at the University of Montenegro, Faculty of Electrical Engineering, Podgorica, Montenegro. She received the BSc and MSc degrees in electronics engineering from the University of Montenegro, in 2003 and 2006, respectively, and is currently working toward her Ph.D. degree at the Center for Telecommunications, Faculty of Electrical Engineering, University of Montenegro. Her main research interests are: wireless communications theory, wireless networks performance improvement, physical layer aspects of wideband cooperative systems.

A New Approach for Increasing Energy Efficiency of OFDM-CDMA System with Pilot Tones

U. Urosevic[1], Z. Veljovic[1] and D. Simunic[2]

[1]Faculty of Electrical Engineering, University of Montenegro, Podgorica, Montenegro; e-mail: {ugljesa, veljovic}@ac.me
[2]Faculty of Electrical Engineering and Computing, University of Zagreb, Zagreb, Croatia; e-mail: dina.simunic@fer.hr

Received 17 November 2012; Accepted 18 December 2012

Abstract

In this paper, incorporation of MISO concept is proposed as a solution for increasing energy efficiency and improving BER performance of OFDM-CDMA downlink system with pilot tone and threshold detection combining (optimum TDC). The new presented system with MISO included uses space-time block coding applied to two, three and four transmit antennas. BER (Bit Error Rate) performance in the case of Ricean frequency selective fading is evaluated for the original system as well as for the one with MISO included. For that reason an adequate simulation model is developed. The results show that the proposed system significantly outperforms the OFDM-CDMA downlink system with pilot tone and optimum TDC. The proposed system provides greater energy efficiency and lower BER in comparison with the original system.

Keywords: MISO, OFDM-CDMA, BER, energy efficiency.

1 Introduction

Future mobile-radio systems should be able to cope with continuously increasing user demands for high bandwidth services and applications. Multi-carrier (MC) techniques represent an attractive solution for achieving high speed data transmissions. Additionally, increased downlink capacities can be accomplished by the combination of MC and code division multiple access (CDMA) [1, 2]. If orthogonal subcarriers are used, this MC-CDMA scheme is known as orthogonal frequency division multiplexing (OFDM-CDMA) [3, 4].

In the case of frequency-selective fading channel, received signals suffer from frequency distortion and thus orthogonality destruction occurs, thereby producing large multi-user interference (MUI). Frequency-domain equalization is necessary at the receiver to restore orthogonality among different users. In [1], OFDM-CDMA downlink scheme with pilot tone and threshold detection combining (optimum TDC) as a combining technique at the receiver is presented. It has been shown that optimum TDC outperforms orthogonal restoring combining (ORC) and controlled equalization combining (CEC). However, implementation of optimum TDC requires computation of optimum weighting coefficients for each SNR, and it is more complex than ORC and CEC.

In the case of high subcarriers offsets due to Doppler spreading, the system presented in [1] shows very poor BER performance. Thus, a solution was proposed in [2] for improving BER performance of the OFDM-CDMA system with pilot tone and optimum TDC, in the case of propagation conditions characterized with high Doppler spread.

Furthermore, in the scheme from [2], in order to improve BER performance transmitted power should be increased or the number of subcarriers used for transmitting single data symbol should be greater, i.e. wider frequency bandwidth is necessary.

Since energy efficiency improvement is one of the most actual problem and research activities on a global level, increasing transmitted power with the goal to decrease BER is not an acceptable solution. Also, radio spectrum available for mobile communications is a scarce resource thus one of the main targets in the design of radio access technologies is to optimize their spectrum efficiency.

In order to achieve better BER performance while providing greater energy efficiency and frequency resources utilization the OFDM-CDMA system from [2] should be improved. Multiple antennas approach is very attractive for that purpose. Multiple antennas technologies are nowadays fully integrated into existing systems, being always foreseen as a promising technology for future communication systems, due to their ability to provide increased data rates, greater energy efficiency, better reliability, improved spectrum efficiency, etc.

In this paper, Multiple-Input Single-Output (MISO) with space-time block coding (STBC) is proposed as a solution for improving energy efficiency and BER performance of the OFDM-CDMA scheme from [2]. It is shown that, with the same transmitted power and spectrum efficiency, the proposed system achieves significantly lower BER than the original scheme. Moreover, the proposed scheme enables better BER performance even with the reduced frequency bandwidth, while retaining the same transmitted power. Also for the same values of BER the proposed system provides representative SNR gains. Besides better energy efficiency, reliability and spectrum efficiency, the presented solution has another advantage since the implementation of a very complex optimum TDC equalization technique is avoided.

The paper is composed in the following order. In Section 2 the proposed MIMO OFDM-CDMA system with STBC is described. Simulation model and results are presented in Section 3. Conclusions are drawn in Section 4.

2 System Model

The transmitter of the proposed MISO OFDM-CDMA downlink transmission system is shown in Figure 1. A single cell system with N users is assumed, with binary transmitted data mapped at the base station according to a chosen modulation scheme. After serial to parallel (S/P) conversion, at every S/P output ($p = 1, \ldots, F$; F is the number of S/P outputs), a space-time block coder is implemented in order to collect a block of M successive symbols which are mapped into a sequence of L consecutive vectors $\mathbf{a}^p[l] = [a_1^p[l] \ldots a_{M_t}^p[l]]^T$, $0 \leq l < L$. The code rate is $R_c = M/L$ and it is assumed that the mobile radio channel remains constant during one code block [6]. In this way, M_t symbols for each l are assigned to M_t STBC outputs, corresponding to transmit antennas. A sequence of pilot tones is implemented at each output of the every space-time block coder and multiplexed with coded data symbols.

Further on, at each STBC output the same signal processing is performed in such a way that after spreading in frequency domain using K-length orthogonal spreading sequence $k = 1, \ldots, K$, $K \geq N$, signals originated from different users are summed. Then, they are multiplied by a long pseudo

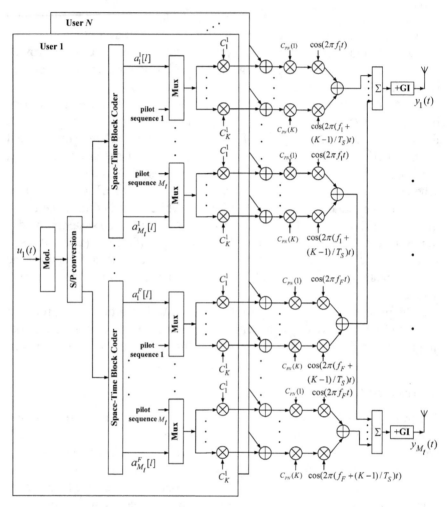

Figure 1 Transmitter of the proposed MISO OFDM-CDMA downlink transmission system.

noise (PN) sequence $C_{PN}(j) = \{-1, 1\}$, $j = 1, 2, \ldots, R$, $R \gg K$, used to randomize the multi-user interference produced by partial destruction of orthogonality due to imperfect frequency equalization. Thereafter, orthogonal subcarriers are modulated.

Orthogonal subcarriers are organized in F groups, each having K sub-carriers mutually separated by $1/T_S$ (T_S is the length of a data symbol). Frequency separation between the subcarriers f_p ($p = 1, \ldots, F$) belonging

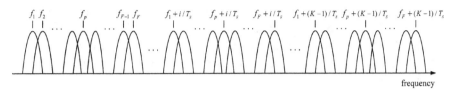

Figure 2 Distribution of subcarriers in the frequency domain.

to different groups, i.e. to different space-time block coders, is $1/FT_S$. Figure 2 shows the subcarriers distribution in the frequency domain. The sum of $F \cdot K$ modulated subcarrier components forms the OFDM-CDMA waveform of length FT_S. This process can be performed using the inverse fast Fourier transform (IFFT). Finally, a guard interval (GI) of length T_{guard} is added to the OFDM-CDMA symbol, to form a resultant OFDM-CDMA symbol of length $T = FT_S + T_{guard}$.

OFDM-CDMA symbols formed in the previously described way are then transmitted by M_t antennas. Thus, a MISO channel is established and for the k-th subcarrier it can be described with the next channel coefficient matrix:

$$\mathbf{H} = [h_1(k) \quad h_2(k) \quad \ldots \quad h_{M_t}(k)], \tag{1}$$

where $h_i(k)$ represents the channel coefficient between the transmit antenna i $(i = 1, \ldots, M_t)$ and the receive antenna.

The output signal $r(k)$ from the receive antenna is given by

$$r(k) = \mathbf{H} \begin{bmatrix} u_1(k) \\ u_2(k) \\ \vdots \\ u_{M_t}(k) \end{bmatrix}. \tag{2}$$

In the relation (2) $u_i(k)$ represents a symbol transmitted from antenna i, while $n(k)$ is an additive white Gaussian noise (AWGN).

3 Simulation Model and Performance Analysis

In order to analyze performance of the proposed MISO OFDM-CDMA system, we have developed an original simulation model. It is based on the assumption that channels between the i-th transmit antenna and the receive antenna are uncorrelated. In that case MIMO channel is modeled with Mt

Single-Input Single-Output (SISO) wireless channels. Propagation conditions are defined with Ricean fading statistics. For the given Ricean channel parameter (K_{dB}), the maximum delay spread (T_{max}), and the rms delay spread (T_{rms}), the simulation model generates one direct and many reflected waves for each channel.

Subcarriers from Figure 1 are realized using the IFFT. In the simulation we have used QPSK modulation, Walsh Hadamard matrix whose rows represent orthogonal spreading sequences for every user, $T_{max} = 150$ ns, $T_{rms} = 150/7$ ns.

As was mentioned in Section 2, space-time block coding is implemented in the proposed system. For the purpose of simulation we have used STBC that can be applied to two, three and four transmit antennas and to an arbitrary number of receive antennas. The space-time codewords arranged in space and time can be described using vector notations [7]:

$$\mathbf{X}_2 = \begin{bmatrix} a_1 & -a_2^* \\ a_2 & a_1^* \end{bmatrix}, \tag{3}$$

$$\mathbf{X}_3 = \frac{1}{\sqrt{12}} \begin{bmatrix} 2a_1 & -2a_2 & \sqrt{2}\,a_3^* & \sqrt{2}\,a_3^* \\ 2a_2 & 2a_1^* & \sqrt{2}\,a_3^* & -\sqrt{2}\,a_3^* \\ \sqrt{2}\,a_3 & \sqrt{2}\,a_3 & -a_1 - a_1^* + a_2 - a_2^* & a_1 - a_1^* + a_2 + a_2^* \end{bmatrix}, \tag{4}$$

$$\mathbf{X}_4 = \frac{1}{\sqrt{16}} \begin{bmatrix} 2a_1 & -2a_2 & \sqrt{2}\,a_3^* & \sqrt{2}\,a_3^* \\ 2a_2 & 2a_1^* & \sqrt{2}\,a_3^* & -\sqrt{2}\,a_3^* \\ \sqrt{2}\,a_3 & \sqrt{2}\,a_3 & -a_1 - a_1^* + a_2 - a_2^* & a_1 - a_1^* + a_2 + a_2^* \\ \sqrt{2}\,a_3 & -\sqrt{2}\,a_3 & -a_1 - a_1^* - a_2 - a_2^* & a_1 - a_1^* - a_2 + a_2^* \end{bmatrix}. \tag{5}$$

$\mathbf{X}_2$, $\mathbf{X}_3$ and $\mathbf{X}_4$ are codewords for systems with two, three and four transmit antennas, respectively. The columns comprise symbols transmitted at a certain time instant, while the rows represent symbols transmitted over a certain antenna.

In analyzing the proposed system, the instantaneous channel estimation for the n-th user at the k-th subcarrier frequency position, between the transmit antenna i and the receive, is taken to be $h_{i,}^n(k)$. Thus, after space-time block decoding, frequency equalization weight factor at the k-th subcarrier frequency position, for the n-th user becomes

$$w_n(k) = \frac{1}{\sum_{i=1}^{M_t} |h_i^n(k)|^2}. \tag{6}$$

Table 1 Simulation parameters.

Modulation	QPSK
System capacity	200 Mb/s
Number of users	64
Subcarrier separation	1/(4TS)
Rice channel parameter (KdB)	4 dB

Other simulation parameters are given in Table 1.

BER performance of the proposed MIMO OFDM-CDMA system is analyzed and compared with OFDM-CDMA system from [2] in different conditions, described with three simulation scenarios. In the first scenario, for both systems we assume the same number of subcarriers that are used for transmitting single data symbol, so the systems have identical spectrum efficiencies. In that manner it is possible to verify if the proposed scheme has better BER performance and energy efficiency while using the same frequency resources. In the second scenario, the spectrum efficiency of the proposed system is examined, so less subcarriers are used in the proposed scheme, i.e. smaller frequency bandwidth is occupied. The aim of this scenario is to investigate if the proposed system can achieve better spectrum efficiency while retaining the lower BER and greater energy efficiency. Finally, in the last scenario BER performance of both systems (with the same number of subcarriers) are compared taking into account Doppler spreading.

We have assumed that 128 subcarriers are used in both systems compared in the first scenario. BER performances over the average received SNR are depicted in Figure 3 for the proposed system (M_t = 2, 3 and 4 transmit antennas) and for the original scheme from [2]. It can be seen from Figure 3 that the proposed system achieves better performance especially in the region of medium and high SNRs. It is obvious that it even provides significantly lower BER when just two antennas at the transmitter side are applied. For example, for the BER value of 10^{-4} the proposed system with Mt=2 transmit antennas attains the SNR gain of around 1.5 dB in comparison with the original scheme, while for the analyzed system with M_t = 3 transmit antennas this SNR gain is about 1.7 dB. In the case of M_t = 4 transmit antennas the proposed system enables SNR gain of about 2 dB.

In the second scenario we have assumed that the proposed scheme uses less subcarriers than the original system. Figure 4 shows BER performances of the proposed system (M_t = 2, 3 and 4 transmit antennas) and of the original scheme from [2]. It has been assumed that 64 subcarriers are used in the scheme including MISO, while the number of subcarriers for the ori-

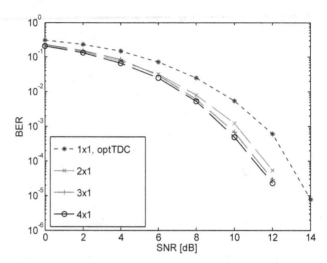

Figure 3 BER performance of the proposed system (M_t = 2, 3 and 4) and of the original scheme from [2].

ginal scheme is 128. From Figure 4 it is clear that, despite using a smaller number of subcarriers, the proposed system has better BER performance. The proposed system even with two times less number of subcarriers is more energy efficient than the original. For example, for the BER value of 10^{-4} the proposed system with M_t = 3 transmit antennas attains the SNR gain of around 0.5 dB in comparison with the original scheme, while for the analyzed system with M_t = 4 transmit antennas this SNR gain is about 1 dB.

Frequency shifts due to Doppler spreading were taken into account in the third scenario. For the various values of the normalized maximum Doppler frequency fdT, BER performances of the proposed and the original scheme from [2], both with 128 subcarriers, are shown in Figure 5. It can be seen from this figure that the proposed scheme obviously outperforms the original scheme from [2]. Also, with increasing frequency shifts, BER increases more rapidly in the original scheme from [2]. For example, for the BER value of 10^{-3} and $f_d T$ = 0.0001, the proposed system with M_t = 2 transmit antennas reduces the required SNR for around 1 dB in comparison with the original scheme, while for the BER value of 10^{-3} and $f_d T$ = 0.0015 this reduction equals around 1.5 dB.

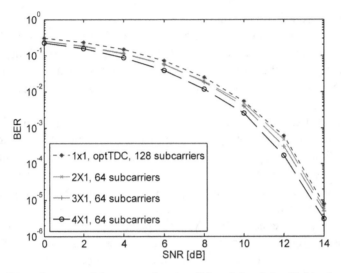

Figure 4 BER performance of the proposed system ($M_t = 2, 3$ and 4) with 64 subcarriers and of the original scheme from [2] with 128 subcarriers.

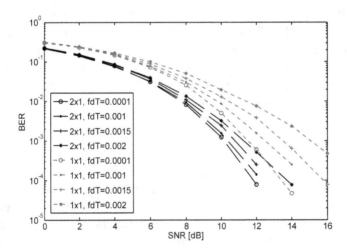

Figure 5 BER performance of the proposed system ($M_t = 2$) and of the original scheme from [2], in the case of Doppler spreading.

4 Conclusion

In this paper, the MISO OFDM-CDMA scheme was proposed as a solution for improving energy efficiency and BER performance of the OFDM-CDMA system with pilot tone and optimum TDC combining. Evaluations and comparisons between the proposed and the original systems were performed using the originally developed simulation model. Three different scenarios were analyzed.

In the first scenario, an equal number of subcarriers is assumed for both systems, with the goal to verify if the proposed scheme can provide better energy efficiency and BER performance while using the same frequency resources. It was shown that the proposed scheme significantly outperforms the original system, even when only two transmit antennas and one receive antenna are applied.

In the second scenario, besides BER performance and energy efficiency, the proposed scheme is analyzed from the point of its spectrum efficiency. Thus, it was shown that with a smaller number of subcarriers, i.e. with higher spectrum efficiency, the proposed system still provides noticeably better BER performance and energy efficiency.

In the last scenario, two schemes were compared taking into account the level of Doppler spreading.

Regarding the obtained results, it is evident that the proposed system can be considered as a good solution for energy efficiency and BER performance improvements of the OFDM-CDMA system with pilot tone. As it was shown, significantly increased energy efficiency, lower BER and frequency resources utilization can be achieved. Besides these improvements, another advantage of the proposed system is in avoiding implementation of the very complex optimum TDC equalization technique.

Acknowledgements

The results presented in this paper are part of the "Improving the concept of MIMO communications in cellular networks of the next generations" research project, funded by Montenegrin Ministry of Science and of the "Performance analysis of OFDM based relay systems" bilateral research project, funded by Montenegrin Ministry of Science and Croatian Ministry of Science, Education and Sports.

References

[1] T. Sao and F. Adachi. Pilot-aided threshold detection combining for OFDM-CDMA downlink transmission in a frequency selective fading channel. IEICE Trans. Commun., E85-B(12): 2816–2826, December 2002.

[2] Z. Veljovic, M. Pejanovic, and I. Radusinovic. Performance analysis of a new OFDM-CDMA system with pilot tone for multimedia communications. IEICE Trans. Commun., E88-B(8): 3480–3483, August 2005.

[3] S. Hara and R. Prasad. Multicarrier Techniques for 4G Mobile Communications. Artech House, 2003.

[4] H. Schulze and C. Lders, Theory and Applications of OFDM and CDMA - Wideband Wireless Communications. Wiley, 2005.

[5] U. Urosevic, Z. Veljovic, and E. Kocan. BER performance of MIMO OFDM-CDMA system in Ricean fading channel. In Proceedings of IEEE EUROCON Conference, Lisbon, Portugal, April 2011.

[6] H. Jafarkhani. Space-Time Coding, Theory and Practice. Cambridge University Press, 2005.

[7] V. Küh., Wireless Communications over MIMO Channels-Applications to CDMA and Multiple Antenna Systems, pp. 283–288. Wiley, 2006.

Biographies

U. Urosevic is a teaching assistant at the University of Montenegro, Faculty of Electrical Engineering in Podgorica, Montenegro. He graduated in 2009 and received his M.Sc. degree in 2011 from the University of Montenegro, Faculty of Electrical Engineering in Podgorica, Montenegro. He is a PhD student now. He is the author or co-author of 15 publications in various conferences and journals. His research areas are wireless communications, MIMO systems, multicarrier systems, 3G and 4G cellular systems, etc.

Z. Veljovic is a professor at the University of Montenegro, Faculty of Electrical Engineering in Podgorica, Montenegro. He graduated in 1992 and received his Ph.D. degree in 2005 from the University of Montenegro, Faculty of Electrical Engineering in Podgorica, Montenegro. He received his M.Sc. degree in 1995 from the University of Belgrade, Faculty of Electrical Engineering in Belgrade, Serbia. He is the author or co-author of 80 papers in various journals and conferences. He is co-author of one publication and two research projects. He is the author or co-author of 10 technical projects, and one is in the framework of EU FP7 program. His research areas include wireless communications, digital communications, digital modulations, multiple access techniques, MIMO systems, satellite communications, etc.

Dina Simunic is a full professor at the University of Zagreb, Faculty of Electrical Engineering and Computing in Zagreb, Croatia. She graduated in 1995 from the University of Technology in Graz, Austria. In 1997 she was a visiting professor in Wandel & Goltermann Research Laboratory in Germany, as well as in Motorola Inc., Florida Corporate Electromagnetics Laboratory, USA, where she worked on measurement techniques, later on applied in IEEE Standard. In 2003 she was a collaborator of USA FDA on scientific project of medical interference. Dr. Simunic is a IEEE Senior Member, and acts as a reviewer of *IEEE Transactions on Microwave Theory and Techniques*, *Biomedical Engineering and Bioelectromagnetics*, *JOSE*, and as a reviewer of many papers on various scientific conferences (e.g., IEEE on Electromagnetic Compatibility). She was a reviewer of Belgian and Dutch Government scientific projects, of the EU FP programs, as well as of COST-ICT and COST-TDP actions. She is author or co-author of approximately 100 publications in various journals and books, as well as her student text for wireless communications, entitled: *Microwave Communications Basics*. She is co-editor of the book *Towards Green ICT*, published in 2010. She is also editor-in-chief of the *Journal of Green Engineering*. Her research work comprises electromagnetic fields dosimetry, wireless communications theory and its various applications (e.g., in intelligent transport systems, body area networks, crisis management, security, green communications). She serves as Chair of the "Standards in Telecommunications" at Croatian Standardization Institute. She servers as a member of the core group of Erasmus Mundus "Mobility for Life".

Comparative Micro Life Cycle Assessment of Physical and Virtual Desktops in a Cloud Computing Network with Consequential, Efficiency, and Rebound Considerations

A.S.G. Andrae

Huawei Technologies Sweden AB, Skalholtsgatan 9, 16494 Kista, Sweden;
e-mail: anders.andrae@huawei.com

Received 30 November 2012; Accepted 28 December 2012

Abstract

The interest in cloud computing is overwhelming and the efficiency/cost gains are thought to be large. Here the potential electricity savings and CO_2e emission are estimated for office usage of physical desktops (PD) and virtual desktops (VD) in a theoretical cloud network. For PD and VD the LCA results are 331–502 and 221–288 kg CO_2e/user/year, respectively. Annually the VD is 18–74% and 20–55% lower than PD for electricity usage and CO_2e emissions, respectively. The normalised numbers for VD for data are 0.30–0.34 kWh electricity/GByte and 0.29–0.38 kg CO_2e/GByte. The allocation of the VD impact to sending and receiving a 500 kByte e-mail was attempted with reasonable results (0.14–1.34 Wh/e-mail). The 2020 global consequential marginal electricity production, dominated by coal and gas, has a slightly higher CO_2e/kWh than the historical global average attributional electricity. Using the consequential marginal electricity does not change any conclusions of the present case study. Electricity efficiency improvements of 5% per year during five years can reduce the VD CO_2e by 11%. Rebound effects for electricity usage off-set somewhat the electricity efficiency improvements. Green powered data centers can reduce VD CO_2e by 20%. Soft clients represented by tablets would reduce VD CO_2e by 1–9% and electricity usage

Journal of Green Engineering, Vol. 3, 193–218.

by 1–16%. Indicatively the share of cloud IP traffic induced CO_2e of global human induced CO_2e is expected to rise between 2012 and 2016.

Keywords: Cloud computing, CO_2e, consequential marginal electricity, data center, desktop, micro life cycle assessment, rebound effect, server, thin client.

Notation

PD	Physical desktop
VD	Virtual Desktop
η	Service Energy Efficiency (kW/Gb/s)
PUE	Power use effectiveness
P	Power usage of Network (kW)
T_n	Total network traffic (Gb/s)
T_s	Bandwidth for studied service, Mb/s
S	Number of annual service sessions
t	Hours per service session
c	Number of years
i	Years
α	efficiency improvement (e.g., electricity)
E_s	Studied service energy usage, kWh
E_{pa}	Annual electricity usage by studied service, kWh
E_{total}	Total electricity usage over c years
E_{pu}	Electricity usage per bit for transferring bits in public core network, J/bit
$E_{pu,VD}$	Electricity usage per bit for transferring bits in public "Network" for VD, J/bit
E_{pr}	Electricity usage per bit for transferring bits in private core network, J/bit
$E_{pr,VD}$	Electricity usage per bit for transferring bits in private "Network" for VD, J/bit
F_o	Overhead factor for redundancy and cooling, around 3
F_u	Overhead factor underutilization of core routers, around 2
O_c	Overprovisioning factor for core routers, around 2
H_c	Hop count factor for core traffic, around 20
P_c	Power usage of Core routers (W)
C_c	Capacity of Core routers, (bits/s)
P_{es}	Power usage of Ethernet switches (W)

C_{es}	Capacity of Ethernet switches, (bits/s)
P_{bg}	Power usage of border gateway routers (W)
C_{bg}	Capacity of border gateway routers, (bits/s)
P_g	Power usage of data center gateway routers (W)
C_g	Capacity of data center gateway routers, (bits/s)
P_{ed}	Power usage of edge routers (W)
C_{ed}	Capacity of edge routers, (bits/s)
P_w	Power usage of wavelength division multiplexing (VDM) equipment (W)
C_w	Capacity of WDM equipment, (bits/s)
P_{ses}	Power usage of small Ethernet switches (W)
C_{ses}	Capacity of small Ethernet switches, (bits/s)
R	Rebound effect (%)
Ex	Expected decrease in electricity usage, kWh
Re	Real decrease in electricity usage, kWh
UPS	Uninterruptable power supply
CC	Climate change
IP	Internet Protocol
ICT	Information communication technology
RMA	Raw material acquisition

1 Introduction

The Information Communication Technology (ICT) sector is growing with a relentless pace. For example, in 2012 Cisco made several worldwide forecasts for 2011 to 2016 on annual basis: mobile cloud traffic will increase from 3.15 ExaByte (10^{18} byte, EB) to 91.2 EB, mobile data traffic from 7.2 to 130 EB, cloud IP traffic from 683 EB to 4.3 Zettabyte (1021 byte, ZB), global data center IP traffic from 1.65 to 6.6 ZB, i.e., cloud IP traffic will be around 2/3 of data center IP traffic in 2016 (Cisco Global Cloud Index 2011–2016).

The ETSI LCA standard for ICT [1] divides the ICT Industry into seven sectors:

1. End-user equipment (e.g., tablets, TVs, mobile phones),
2. Customer Premise Equipment (e.g. gateways modems, set top boxes),
3. Access network (e.g. xDSL, BTS),
4. Core & control network (e.g., ethernet switches, core routers),
5. Data Centers (e.g. servers, storages),
6. Data transport (e.g., metro switches), and

7. Service Provider activities (e.g., software development).

Somavat and Namboodiri [2] estimated that the ICT segment (ETSI Sectors 1, 3, 4, 5) used 6% of the global electricity in 2010.

Lambert et al. [3] estimated that the use stage worldwide electricity usage of communication networks, ETSI sectors 2, 3, and 4, will exceed 350 TWh in 2012, around 1.8% of global total usage. IEA estimated use stage around 900 TWh for sectors 1 to 4 (International Energy Agency, 2009). In 2012 this is around 4% of the total global electricity usage. Corcoran [4] compiled sectors 1 to 6 and concluded that ICT use stage electricity usage is 9% of global total in 2012 growing to more than 20%(!) in 2016. These usages corresponds roughly to 3.8% and 10.1% of human induced CO_2e emissions assuming an annual 0.5% increase of CO_2e emissions and 31 billion tonnes released in 2012.

Note that these estimations have excluded the raw material acquisition, production, distribution, and end-of-life treatment of the ICT Equipment which could be very significant, especially for end-user equipment. Anyway, the overall electricity usage driven by the seven ICT sectors will grow fast in the next years, especially for sectors 1, 2, and 5.

New cost effective platforms such as cloud computing are emerging where the benefit is delivered as services rather than products [5–8]. Anyway, the ICT Sector should also strive for as green delivery as possible. A truly green delivery is always characterized by low economic cost and high extended exergy efficiency, or even net exergy generation.

In this context a small scale theoretical case study investigation was done to start to understand some of the environmental implications of cloud computing. Maga et al. [9] presented a micro attributional LCA where desktop based computing was compared to thin client based computing. The results were a 70% advantage for thin client based computing.

The news value here compared to Maga et al.'s [9] is that system boundaries are wider, moreover the present sensitivity analyses are much more extensive, and in addition an ICT Service has been estimated.

The problems addressed: What are the implications for electricity and CO_2e for a comparative micro LCA study of physical desktop (PD) and virtual desktop based computing (VD) in cloud networks? What is the approximate electricity usage and CO_2 emissions associated with sending and receiving a 500 kByte email using VD?

2 Materials and Methods

The present research is based on the concept of micro attributional LCA. Most LCAs performed to date are micro attributional LCAs, nevertheless micro consequential LCAs are also becoming more common.

The added value of the present paper is the first case study to the author's knowledge which uses micro LCA on an extensive desktop cloud network and partially include electricity efficiency and rebound effects. Moreover the LCA result is applied to estimate a result to for a specific cloud service supplied by the cloud network.

2.1 LCA Basics

LCA is a standardized method [1] for making model based estimations of the environmental exchanges associated with technology functions. Usually LCAs are performed on small product systems and the results are not generally applicable to global conclusions, even though attempts have been made to scale up micro LCAs to estimate global consequences, e.g., for solders [10]. Therefore the prefix micro is added to this LCA. In fact Dandres et al. [11] recently proposed the conceptual model of macro LCA which will mean a paradigm shift for LCA application. The conceptual LCA models are attributional LCA and consequential LCA. Consequential LCA is preferable when the difference between attributional LCA results is small and the rebound effects are thought to be large.

Micro LCA studies are required to follow four main steps:

1. Goal & Scope definition,
2. Inventory analysis,
3. Impact assessment of the inventory, and
4. Interpretation of the impact assessment.

The goal & scope section includes the definition of the functional unit which must express the function of the system in a measurable unit. The f.u. shall moreover attempt to answer the questions: What? Where? When? How well?

The impact assessment is usually done with so called mid-point and/or end-point valuation.

The present paper will report climate change and electricity results and exhibit limitations and strengths of micro LCAs applied to complex ICT systems.

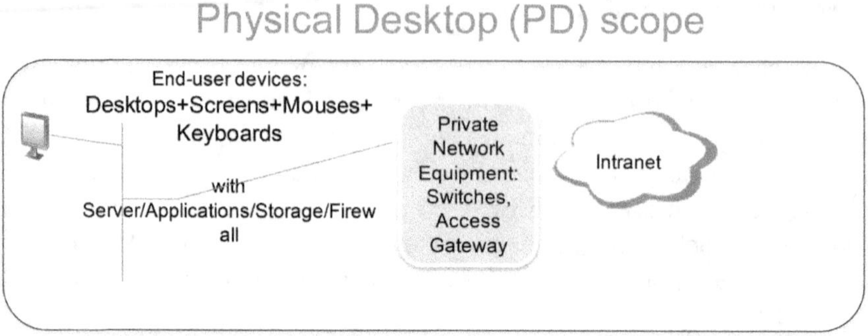

Physical Desktop (PD) scope

Figure 1 Studied product system of LCA study for a physical desktop.

3 Goal & Scope

The studied product systems (SPS) are shown in Figures 1 and 2. Global average historical electricity production and usage is also included wherever electricity is used. Later the global marginal electricity will be used in the sensitivity analyses.

3.1 Functional Unit

For the micro LCA the functional unit (f.u.) is *the annual usage of the possibility to perform office computing daily work in a "global" office by 488 users*, and the system boundaries are from cradle-to-grave. Indeed this f.u. is intentionally broad to capture "all" functions of office computer work. Strictly there could be situations where PD and VD provide different functions. However, the office usage typically includes sending/receiving emails, file creating, and software program handling. All tasks are either done locally on desktops or on data center servers over which virtual machines are run.

In Table 1 a summary of SPS, cut-off and excluded building blocks is shown. The cut-off criterion is 5% addition to the first iteration of GWP 100 or electricity usage.

Typical for ICT networks is that the lifetime of system parts vary [12] and this has to be handled when expressing the result annually as required by the recent ETSI LCA standard [1].

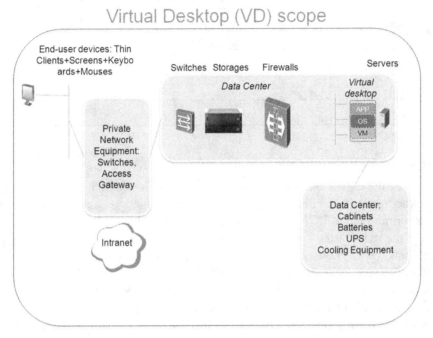

Figure 2 Studied product system of LCA study for a virtual desktop.

Table 1 Summary of SPS and cut-off for physical and virtual desktops in global office.

Computing type → ↓ System	PD	VD
SPS	RMA+Production+Assembly+Distribution of hardware to Use, Use of hardware, and End-of-life treatment of hardware.	RMA+Production+Assembly+Distribution of hardware to Use, Use of hardware, and End-of-life treatment of hardware.
Cut-off from SPS	Service provider activities, Distribution of hardware, Installation, De-installation, End-of-life treatment of hardware, Cables, Support activities	Service provider activities, Distribution of hardware, Installation, Deinstallation, End-of-life treatment of hardware, Cables, Waste heat recovery in data centers, Support activities
Excluded from the start	Maintenance & repair, Office area furniture &buildings, Re-use of hardware	Maintenance & repair, Office area furniture&buildings, Data center building, Re-use of hardware

Table 2 Summary of life cycle inventory for PD and VD per functional unit (f.u.).

Substance	Unit	Indicative Uncertainty	PD	VD
CO_2	kg	±20%	189,900	114,500
CH_4	kg	±50%	530	315
N_2O	kg	±30%	6.19	3.81

3.2 Inventory Analysis

It is assumed that "unspecified electronics" emit around 30 kg CO_2e/kg from cradle-to-use [13]. For the time being this approximation could be used when performing micro LCAs of large ICT systems for which many parameters except the mass of the "electronics" inside individual ICT Equipment/Devices are unknown. These approximations would currently make the overall Network/Service LCA non compliant with ETSI LCA standard [1]. In spite of this, "unspecified electronics" can still be used to estimate the magnitude of the hardware production compared to use stage in complex ICT Networks consisting of many Equipment. In Table 2 a summary of the inventory analysis for the present micro LCA is shown.

The present micro LCA model for VD consists of 16 main modules created from internal investigations founded in internal and external literature (Table 3). The quality of the inventory data is here considered high enough for the purpose of estimating PD and VD. The lack of data quality is paid attention to as uncertainties, e.g. lifetime uncertainty, are taken into account. For the hardware in Table 3 the CO_2e per unit is for cradle-to-use whereas "Electricity, global average" is for both upstream and use stages. Tablet is used only for sensitivity analysis for the Soft Client scenario for VD.

Table 3 shows that the relative share of screen production of the total PD and VD results can be expected to be relatively large. Table 4 shows the electricity usages by end-user and data center equipment.

Lenovo's Energy Calculator was used for estimating Screen and Desktop use stage electricity usages. The ancillary electricity used annually by the data center was estimated by assuming a typical PUE of 1.7 [20] which was multiplied with the sum of annual Equipment (numbers 7–10 in Table 4) usages of data center equipment: (PUE-1)×31,534 kWh = 22,100 kWh/year. There are many other green performance indicators for data centers [14], however, their application is beyond the purpose and scope of this paper.

Table 3 Summary of life cycle CO_2e emissions of main LCA modules for physical and virtual desktops.

Name	Unit	kg CO_2e per Unit	Uncertainty (±%)	Pieces used by PD per f.u.	Pieces used by VD per f.u.
1. Keyboards	piece	26	35	163	163
2. Mouses	piece	5	25	163	163
3. Thin Clients	piece	32.5	30	-	98
4. Screens	piece	322	20	163	163
5. Desktops	piece	240	40	163	-
6. Electricity, global (average retrospective)	MWh	627	10	141.9	120.8
7. Blade servers	piece	564	30	–	4
8. Storages	piece	2,540	30	–	0.4
9. Switches	piece	282	30	–	0.8
10. Firewalls	piece	282	30	–	0.4
11. Batteries	piece	36	15	–	10
12. Cabinets	piece	639	15	–	0.2
13. UPS	piece	2,900-11,900	10	–	0.05
14. Air conditioners	piece	6,350-7,430	10	–	0.1
15. Network Switches	piece	282	30	12	12
16. Network Access Gateways	piece	141	30	0.2	0.2
17. Tablets	piece	22	30	–	244
18. Distribution	ton	0.57	60	2.94tonnes	1.33tonnes

Table 4 Summary of annual electricity usages for main LCA modules for physical and virtual desktops.

Name	Unit	Amount per year per piece	Uncertainty (±%)	Pieces used at the same time per f.u.
3. Thin Clients	kWh	35.7	10	488
4. Screens	kWh	47	10	488
5. Desktops	kWh	286	50	488
7. Blade servers	kWh	800	30	20
8. Storages	kWh	5,560	10	2
9. Switches	kWh	777	30	4
10. Firewalls	kWh	653	30	2
15. Network Switches	kWh	777	30	12
16. Network Access Gateways	kWh	574	30	1
Data center ancillary electricity	kWh	22,100	10	1
17. Tablets	kWh	6.44	10	488

4 Results

4.1 Impact Assessment

In Figures 3–8 some GWP100 and electricity usage results are shown for VD and PD, respectively.

Figures 3–8 indicate that production of End-use devices is significant.

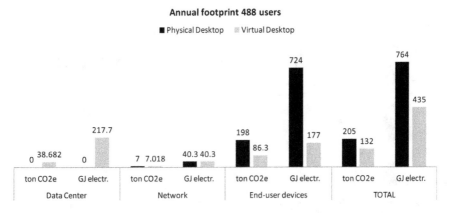

Figure 3 CO_2e and electricity footprint for PD and VD.

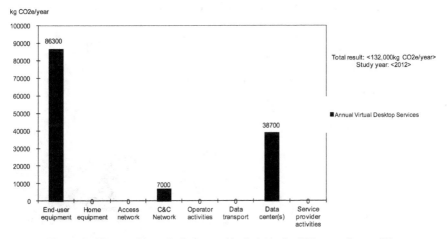

Figure 4 Climate change indicator result diagram for VD according to [1].

5 Allocation of the VD LCA Result to an ICT Service – E-mail

In this section the previous electricity and power results (Figure 3) for VD will be used to estimate the CO_2e and electricity for a specific ICT service delivered by VD: sending and receiving e-mails. Around 2 million e-mails are handled annually each with an average size of 500 kByte.

The power usage for VD is approximately $(120,833kWh)/(16 hours/day \times 365) = 20.69$ kW. Several researchers have proposed different calculation

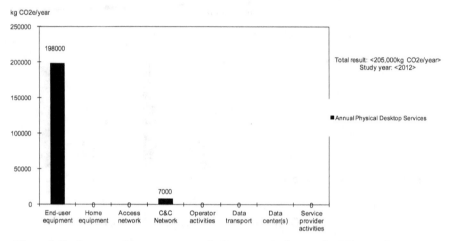

Figure 5 Environmental impact category indicator result diagram for PD according to [1].

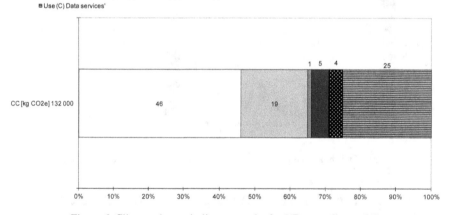

Figure 6 Climate change indicator results for VD according to [1].

schemes for estimating the electricity or power used per ICT Service [6, 15, 16]. For example, Williams and Tang [17] estimated around 0.17 Wh electricity and 0.17 g CO_2/Mbyte for downloading a 1.63 Gbyte software program in the UK.

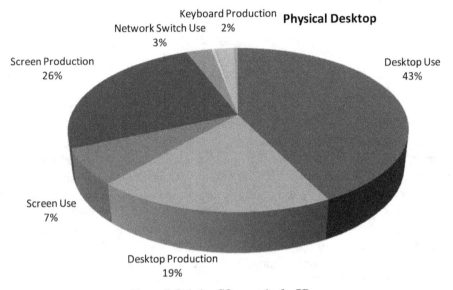

Figure 7 Relative CO_2e results for PD.

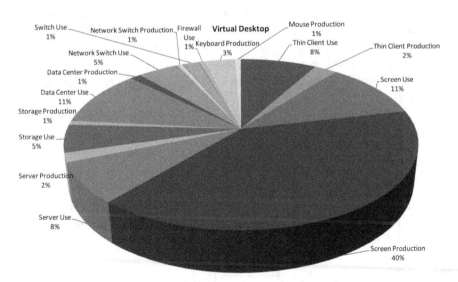

Figure 8 Relative CO_2e result for VD.

The Centre for Energy Efficient Telecommunications (CEET) formulae (Eqs. 1–3) [15] are used below.

$$\eta = \frac{P}{T_n} \tag{1}$$

$$E_s = \eta \times T_s \times t \tag{2}$$

$$E_{pa} = E_s \times S \tag{3}$$

$P = 20.69$ kW, $T_n = 0.0976$ Gb/s total average network traffic, $T_s = 6$ Mb/s average bandwidth email service, $S = 2$ million e-mails sent/received per year, $t = 1.4$ seconds to complete the task. Hence, $\eta = 211.92$ [kW]/[Gb/s], $2 \times E_s = 0.99$ Wh/e-mail, and $2 \times E_{pa} = 1.98$ MWh/year, i.e., less than 1% of VD annual electricity usage can be attributed to e-mails. The uncertainties of P, T_n, T_s and t lead to a spread of at least 0.14–1.34 Wh/email. The E_s and E_{pa} values had to be doubled to account for the receiving of the e-mails. The Wh/e-mail is strongly dependent on T_n, T_s, and t.

CEET also proposed to integrate the electricity usage over time and compare the result with Top Runner Equipment to obtain a so called green service index rating (GIR). The GIR estimation for VD e-mail is beyond the scope of this paper.

A method for estimating the electricity usage for the "Network" and "Data Center" of transporting a bit from the data center to the user via public and private networks is presented by Mouftah and Kantarci [6]. For the number of hops in the public network, H_c, 20 was used. Moreover, typical energy efficiency values (W/(Gbits/s)), e.g., from Van Heddeghem et al. [18] and Chowdhury [19] were used for all kinds of equipment in Eqs. (4–7), e.g., $\eta_{ses} = P_{ses}/C_{ses} = 1.3$ W/[Gb/s].

For the private networks, which resemble the private intranet for VD, the following formula was given:

$$E_{pr} = F_o \times F_u \left(\frac{P_{ses}}{C_{ses}} + 3 \times \frac{P_{es}}{C_{es}} + \frac{P_g}{C_g} \right) \tag{4}$$

As Eq. (4) includes regular ethernet switches and gateway routers located inside the data center, two ethernet switches are removed to obtain the "Network" share of VD:

$$E_{pr,VD} = F_o \times F_u \left(\frac{P_{ses}}{C_{ses}} + \frac{P_{es}}{C_{es}} \right)$$

$$= \{\eta_{ses} = 1.3 \ (\text{W/Gb/s}), \eta_{es} = 8\} = 5.58 \times 10^{-8} \ \text{J/bit} \tag{5}$$

For public networks, which resemble public external Internet, the following formula was given:

$$E_{pu} = F_o \times F_u \left(3 \times \frac{P_{es}}{C_{es}} + \frac{P_{bg}}{C_{bg}} + \frac{P_g}{C_g} + 2 \times \frac{P_{ed}}{C_{ed}} \right.$$

$$\left. + O_c \times H_c \times \frac{P_c}{C_c} + (H_c - 1) \times \frac{P_W}{2C_W} \right)$$

$$= \{\eta_{es} = 8 \text{ (W/Gb/s)}, \eta_{bg} = 80, \eta_g = 137, \eta_{ed} = 26, \eta_c = 12, \eta_w = 3\}$$

$$= E_{pu} = 4.81 \times 10^{-6} \text{ J/bit} \tag{6}$$

Exluding equipment located in the data center,

$$E_{pu,VD} = F_o \times F_u \left(\frac{P_{es}}{C_{es}} + \frac{P_{bg}}{C_{bg}} + 2 \times \frac{P_{ed}}{C_{ed}} \right.$$

$$\left. + O_c \times H_c \times \frac{P_c}{C_c} + (H_c - 1) \times \frac{P_W}{2C_W} \right)$$

$$= E_{pu,VD} = 3.89 \times 10^{-6} \text{ J/bit} \tag{7}$$

The public internet network contains more types of equipment than the private intranet and therefore uses more power. The J/bit values obtained from Eqs. (4–7) for public and private networks are highly dependent on the energy efficiencies of the individual equipment. Moreover, Eqs. (4–7) do not include electricity used in raw material acquisition and production.

6 Discussion

The most important part of an LCA is the interpretation which includes contribution, uncertainty, and sensitivity analyses in which the robustness of the results is tested.

According to contribution analysis (see Figures 3–8) the most important phase is Screen production. Specifically for VD the most contributing processes are screen production and data center cooling.

The Network part ("transmission", "C&C Network", "intranet", "transport network") of the PD and VD systems was estimated to 7,000 kg CO_2e per year. Using 5.58×10^{-8} J electricity/bit from Eq. (5) and the annual bits used by 488 users (around 3.22×10^{15} bits) the CO_2 from Network equipment for PD and VD both are 31,000 kg CO_2e per year. Yet, all bits are not transmitted over the data center as only actual actions done by the user are transported.

Scharnhorst [20] estimated in 2006 the emission cost of transmitting data from a mobile phone via UMTS network to 5×10^{-8} kgCO$_2$e/bit, i.e., 429 kgCO$_2$e/GByte. Although not comparable (different functional unit and system boundary) the present cloud VD network uses around 0.29–0.38 kgCO$_2$e/GByte. This indicates that the bit based energy efficiency within the ICT Sector has been dramatically improved.

Anyway, using the present results of 0.29–0.38 kg CO$_2$e/GByte and Cisco's 2012 and 2016 estimations for IP traffic some forecasting follows. Cloud IP traffic CO$_2$e will rise from 0.78 to 4.9% of global human induced emissions, and data center IP traffic from 1.9 to 7.5%.

If 6 billion users worldwide use VD, their share of the annual global human induced emissions would be around 5%.

6.1 Interpretation – Uncertainty

The uncertainty analysis in LCA should mainly investigate how the precision of used data influence the spread of the final score. The difference between the PD and VD systems was shown to be enough to draw conclusions as the LCA tool SimaPro quantifies the process correlation. On the other hand, the uncertainties of GWP100 indices were not included here. The mean CO$_2$e score for VD is 132 tonnes (95% confidence interval, 108–141), for PD 205 tonnes (162–245), and as shown in Figure 9, PD-VD 73 tonnes (41–116).

6.2 Interpretation – Sensitivity

The processes and activities cut-off from SPS are modelled with less detail and data quality. These processes (Table 1) should be added to investigate their importance and to determine if more data collection is needed.

6.2.1 Distribution & End-of-Life (Inclusion)

Several LCAs for ICT systems have shown that Climate Change scores are not sensitive for end-of-life CO$_2$e. On the positive side metals (Au, Ag, Cu, Pt, Pd) are partly recovered still there are also several truck transports involved. Considering the relatively high production impacts for displays, re-use of Screens could be promoted. Preliminary calculations show that re-using 50% of the Screens for four more years would decrease the CO$_2$e for PD and VD by some 10%. However, this estimation would need a lot more detail.

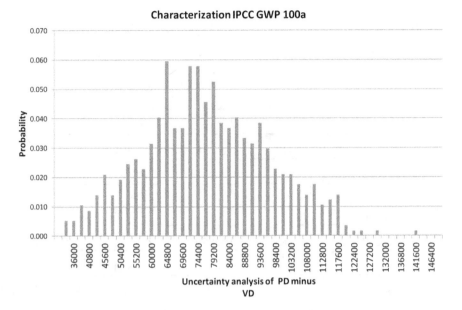

Figure 9 Result (kg) of a 600 runs Monte Carlo Simulation for PD-VD.

The final distribution of ICT Equipment is often done by ships which contribute relatively low CO_2e emissions often making detailed distribution analysis superfluous. Arguably, several truck transports are always attached to the ship transports which increase the transport impacts. Adding distribution impacts (Table 3) to all hardware increased the total score by 1–2%. It seems that ICT Services LCAs are not sensitive to transport impacts. In the end the relevance of distribution depends on the transport modes used for the specific ICT Equipment on which the Service delivery is based.

6.2.2 Soft Client instead of Thin Client
In this sensitivity check a tablet (0.58 kg, 3,250 mAh) is used instead of the thin client (1.1 kg). The tablet could emit 13–26 whereas the thin client emits in the vicinity of 26–38 kg CO_2e/year.

Overall the VD CO_2e score and electricity usage could decrease by 0.25–9% and 0.4–16%, respectively, if tablets replace thin clients. Possibly air transports of tablets could increase their impact, but will not increase significantly.

Table 5 Global gross generation of electricity by source (International Atomic Energy Agency, 2012).

Source of electricity	Gross generation in 2000 (TWh)	Gross generation in 2010 (TWh)	Predicted Gross generation in 2020 (TWh)
Coal	5,989	7,143	9,075
Oil	1,241	1,348	1,371
Gas	2,676	4,947	7,696
Hydrogen-fuel cells	0	0	15
Nuclear	2,586	2,889	2,758
Hydro	2,650	3,188	3,800
Other renewables	249	521	863
TOTAL	**15,391**	**20,037**	**25,578**

6.2.3 Future Long-Term Marginal Electricity Mix Instead of Average Retrospective Electricity Mix

The global electricity usage will increase despite the reduction induced by the "micro" shift from PD to VD. It has started to become normal practise in LCA to include marginal electricity as it is more realistically reflects how investments are carried out today, and in the future and how that is related to the demand for electricity. Actually average electricity should not be used for decision making which involve changed electricity usage as the average does not describe the magnitude of emission changes. Here the effect on the long-term global electricity mix of increased electricity usage is estimated.

In this scenario the so called consequential future electricity mix is used instead of average historical mix (attributional) worldwide. In Tables 5 and 6 the model for establishing consequential electricities suggested by Merciai et al. [21] was used.

This estimation underlines that long-term coal and gas will dominate the consequential future marginal global electricity mix suggesting higher CO_2e emissions per kWh than attributional mix. The consequential historical mix,

Table 6 Global gross generation of electricity by source.

Source of electricity	Consequential future		Consequential historical		Attributional 2009 (International Energy Agency, 2009)
	Predicted change (TWh), 2010-2020	Applied mix	Change (TWh), 2000-2010	Applied mix	Applied mix
Coal	1,932	0.349	1,154	0.248	0.407
Oil	23	0.004	107	0.023	0.054
Gas	2,749	0.496	2,271	0.489	0.211
Hydrogen-fuel cells	15	0.003	0	0.000	0
Nuclear	-131	0.000	303	0.065	0.134
Hydro	612	0.110	538	0.116	0.161
Other renewables	342	0.062	272	0.059	0.03
TOTAL	**5,542**	**1**	**4,645**	**1**	**1**

lower than attributional mix and based on measured data, can be used for sensitivity analyses when the predictions for the future marginal is considered too uncertain. Additionally, the predictions in Table 6 do not include annual CO_2 abatement efficiency improvements for coal, oil, and gas. Moreover (low CO_2e) nuclear is not part of the consequential future marginal mix as nuclear electricity vendors do not respond to changes in demand (according to this model).

The average historical electricity mix (attributional) used previously in Figures 3–8 is replaced by the consequential future mix. The relative and absolute impacts per f.u. are shown in Figure 10.

The effect of using consequential marginal electricity mixes could be larger on individual region or national level than on global level.

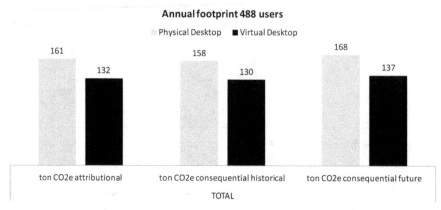

Figure 10 The effect on CO_2e emissions for PD and VD of using consequential future electricity.

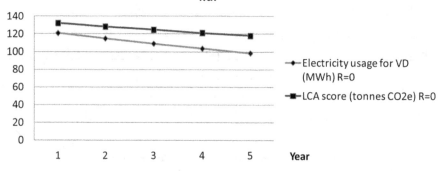

Figure 11 The effect of 5% annual electricity efficiency improvement on VD CO_2e score.

6.2.4 Electricity Efficiency Improvements over Time

Figure 11 shows the effect on electricity usage and CO_2e score for VD if the electricity efficiency is improved 5% per year. CEET [15] proposed Eq. (8) to quantify the electricity usage over time.

$$E_{\text{total}} = \sum_{i=1}^{c}(E_{pa} \times (1 - \alpha)^i) \qquad (8)$$

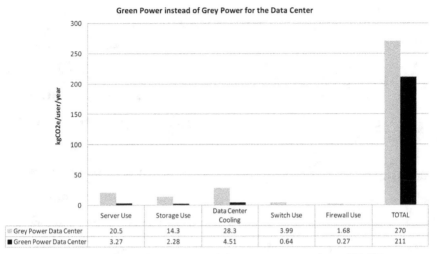

	Server Use	Storage Use	Data Center Cooling	Switch Use	Firewall Use	TOTAL
Grey Power Data Center	20.5	14.3	28.3	3.99	1.68	270
Green Power Data Center	3.27	2.28	4.51	0.64	0.27	211

Figure 12 The effect of green powered data center on annual user VD CO_2e.

6.2.5 Green Powered Data Centers

In an ideal world the data centers worldwide could locally be powered by so called green power [14, 22–25]. Figure 12 shows the potential saving (22%) for the present micro LCA system for VD expressed per annual user should the data center use green power.

6.2.6 Rebound Effect

Simply stated the rebound effect is that efficiency improvements have not led to expected reductions of usage. For example, Rostedt [26] found that Sweden's population consumed a similar proportion energy of Gross Domestic Product in 2000 as they did in the early 1900s, despite several energy efficiency improvements during the same period. The reason is a rebound effect on energy-intensive products. The rebound effect (%) is defined as

$$R = \frac{Ex - Re}{Ex} \tag{9}$$

A rebound effect of 40% means that, e.g., the electricity usage decrease 60% of the expected decrease, $(100 - 40)/100\% = 60\%$. Eq. (8) is modified into Eq. (10) to include the rebound effect, R:

$$E_{\text{total}} = \sum_{i=1}^{c} (E_{pa} \times (1 - \alpha \times R)^i) \tag{10}$$

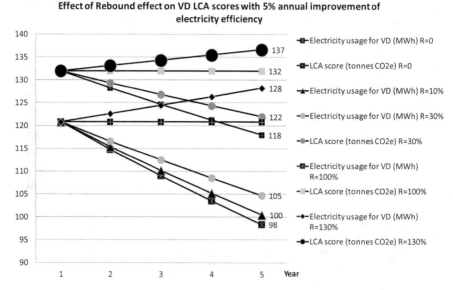

Figure 13 The rebound effect for annual electricity usage on VD CO$_2$e.

A rebound effect of 130% means that the electricity usage will increase despite efficiency improvements, i.e., a "backfire" will happen. Historically this has only happened when totally new markets have appeared, such as the discovery of electricity in the beginning of the 20th century. It could be argued that cloud computing is paving the way for totally new markets. Applied to VD, this phenomenon is shown in Figure 13 for electricity rebound effects.

Figure 14 shows the cumulative effect of Eq. (10) for electricity usage and CO$_2$e of different degrees of rebound. E.g., over five years, with a rebound effect of 100%, VD uses 10% more electricity and emit 6% more CO$_2$e than rebound effect 0%.

7 Conclusions

For the first time physical desktops and virtual desktops have been simultaneously compared in an extensive cloud network by micro LCA with a functional unit of annual usage of the possibility to perform office computing daily work in a global office by 488 users. The results of this analysis show that VD can reduce the CO$_2$e emissions and electricity usage by 18–74% and 20–55%, respectively. Moreover uncertainty and several sensitivity analyses

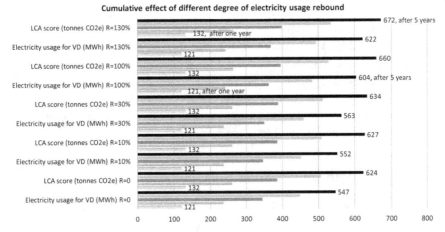

Figure 14 The cumulative effect on electricity usage and CO_2e for VD.

showed the relative result to be robust. The rebound effect could more or less off-set the absolute results for PD and VD both. Still it is highly likely that VD is more electricity efficient than PD.

8 Recommendations and Perspectives

For VD the following design recommendations are valid: use screens as long as possible, optimize data center cooling or switch from grey power to re-newable power, reduce screen power, reduce thin client power, reduce server power. It is evident that comparative micro LCAs do not tell enough about the total global consequences of the technology shift they analyse. Approaches such a macro LCA [10] is likely needed to get nearer to correctly including the rebound effects. Each micro LCA has unique conditions which affect the validity of the final results, both for system and services. Therefore a new micro LCA has to be done for each new system, especially for ICT systems.

From a Service LCA perspective, it was beyond the scope of the case study to find out if VD is also more effective (accomplished work compared to planned target) in specific working situations than PD. The procedure for individual ICT Service calculation has started and it remains to be seen if the uncertainties will be too large to compare and put energy ratings on unique ICT Services in a meaningful way. Normally, data inputs are needed which depend on the situation at hand involving more or less complicated assump-tions. Regarding the robustness of estimations for specific ICT Services, such as video conference, CEET [15] estimated 42.25 Wh electricity/hour (around 26 gCO_2e/h) including only the use stage whereas three other studies report 2 to 7.5 kg/h [27]. This shows that the background information (calculation model and system boundary) is crucial to understand the meaningfulness of the values. Standardization [27] will be able to address most issues regarding understanding of assumptions.

A next step is to make dynamic global assessments of cloud computing platforms including laptops and other equipment.

Acknowledgement

Support from Huawei Technologies CO. Ltd., Xie Yongming is gratefully acknowledged.

References

[1] European Telecommunications Standards Institute. ETSI TS 103 199 V1.1.1 (2011-11) Environmental Engineering (EE); Life Cycle Assessment (LCA) of ICT equipment, networks and services; General methodology and common requirements. 2011. URL: http://www.etsi.org/deliver/etsi_ts/103100_103199/103199/01.01.01_60/ ts_103199v010101p.pdf. Accessed: December 21, 2012.

[2] P. Somavat and V. Namboodiri. Energy consumption of personal computing including portable communication devices. Journal of Green Engineering, 1(4):447–475, 2011.

[3] S. Lambert, W. Van Heddeghem, Willem Vereecken, B. Lannoo, D. Colle, and M. Pickavet. Worldwide electricity consumption of communication networks. Optics Express, 20(26):B513–B524, 2012.

[4] P.M. Corcoran. Cloud computing and consumer electronics: A perfect match or a hidden storm? IEEE Consumer Electronics Magazine, 1(2):14–19, 2012.

[5] H. Pandey. Present scenario analysis of green computing approach in the world of information technology. Undergraduate Academic Research Journal (UARJ), 1(2):1–5, 2012.

[6] H.T. Mouftah and B. Kantarci. Energy-efficient cloud computing – A green migration of traditional IT. In Mohammad S. Obaidat, Alagan Anpalagan, and Isaac Woungang (Eds.), Handbook of Green Communications, 1st ed., pp. 295–329. Academic Press, Elsevier, Oxford, UK, 2013.

[7] P. Sharma and A.J. Singh. A green-cloud network scenario: Towards energy efficient cloud computing. International Journal of Advanced Research in Computer Science and Software Engineering, 2(10):154–158, 2012.

[8] S.J. Prakash, K. Subramanyam, and U.D.S.V. Prasad. Towards energy efficiency of green computing based on virtualization. International Journal of Emerging Trends in Engineering and Development, 7(2):415–423, 2012.

[9] D. Maga, M. Hiebel, and C. Knermann. Comparison of two ICT solutions: Desktop PC versus thin client computing. International Journal of Life Cycle Assessment, DOI 10.1007/s11367-012-0499-3, 2012.

[10] A.S.G. Andrae. Global Life Cycle Impact Assessments of Material Shifts: The Example of a Lead-Free Electronics Industry. Springer Verlag, London, 2009.

[11] T. Dandres, C. Gaudreault, P.T. Seco, and R. Samson. Macroanalysis of the economic and environmental impacts of a 2005–2025 European Union bioenergy policy using the GTAP model and life cycle assessment. Renewable and Sustainable Energy Reviews, 16(2):1180–1192, 2012.

[12] J. Bonvoisin, A. Lelah, F. Mathieux, and D. Brissaud. An environmental assessment method for wireless sensor networks. Journal of Cleaner Production, 33:145–154, 2012.

[13] A.S.G. Andrae and O. Andersen. Life cycle assessments of consumer electronics – Are they consistent? International Journal of Life Cycle Assessment, 15(8):827–836, 2010.

[14] S.S. Mahmoud and I. Ahmad. Green performance indicators for energy aware IT systems: Survey and assessment. Journal of Green Engineering, 3(1):33–69, 2012.

[15] C.A. Chan, E. Wong, A. Nirmalathas, A.F. Gygax, C. Leckie, and D.C. Kilper. Towards and energy rating system for telecommunications. Telecommunications Journal of Australia, 62(5), 2012.

[16] Sonia, S. Pal. Analysis of energy consumption in different types of networks for cloud environment. International Journal of Advanced Research in Computer Science and Software Engineering, 2(2), 2012.

[17] D.R. Williams and Y. Tang. Methodology to model the energy and greenhouse gas emissions of electronic software distributions. Environ. Sci. Technol., 46(2):1087–1095, 2012.

[18] W. Van Heddeghem, F. Idzikowski, W. Vereecken, D. Colle, M. Pickavet, and P. Demeester. Power consumption modeling in optical multilayer networks. Photonic Network Communications, 24(2):86–102, October 2012.

[19] P. Chowdhury. Energy-efficient next-generation networks (E2NGN). Ph.D. Thesis, Computer Science, University of California Davis, US, p. 18.

[20] W. Scharnhorst, L.M. Hilty, and O. Jolliet. Life cycle assessment of second generation (2G) and third generation (3G) mobile phone networks. Environment International, 32(5):656–675, 2006.

[21] S. Merciai, H. Schmidt, and R. Dalgaard. Inventory of country specific electricity in LCA – India. Inventory report v2. 2.0 LCA consultants, Aalborg, 2011. http://www.lca-net.com/projects/electricity_in_lca/

[22] G. Koutitas and P. Demestichas. Challenges for energy efficiency in local and regional data centers. Journal of Green Engineering, 1(1):1–32, 2010.

[23] G. Koutitas and P. Demestichas. A review of energy efficiency in telecommunication networks. Telfor Journal, 2(1):2–7, 2010.

[24] M. C. Lucchetti, R. Merli, and C. Ippolito. A new environmental challenge: How to improve data center energy efficiency. J. Commodity Sci. Technol. Quality, 49(3):173–190, 2010.

[25] K. Kumon. Overview of next generation green data center. Fujitsu Sci. Tech. J., 48(2):177–183, 2012.

[26] J. Rostedt. Energy efficiency, energy consumption, growth, climate issue – Do energy efficiencies lead to reduced energy consumption. Master Thesis No. 740, Department of Economy, Swedish University of Agricultural Sciences, 2012 [in Swedish].

[27] A.S.G. Andrae. European LCA standardization of ICT: Equipment, networks, and services. In Matthias Finkbeiner (Ed.), Towards Life Cycle Sustainability Management, 1st ed., pp. 483–493. Springer, Berlin, 2011.

Biography

Anders S.G. Andrae received the M.Sc. degree in chemical engineering from the Royal Institute of Technology, Stockholm, Sweden, in 1997, and the Ph.D. degree in electronics production from Chalmers University of Technology, Gothenburg, Sweden, in 2005. He worked for Ericsson with LCA between 1997 and 2001. Between 2006 and 2008 he carried out post-doctoral studies at the National Institute of Advanced Industrial Science and Technology (AIST), Tsukuba, Japan. His specialty is the application of sustainability assessment methodologies to ICT solutions from cradle-to-grave. Dr. Andrae is recently the Editor of European Telecommunications Standards Institute (ETSI) first LCA standard for ICT. He has previously published three books, three theses, 19 conference papers, and 14 peer-reviewed journal papers. Since 2008 Dr. Andrae has been with Huawei Technologies in Sweden as Senior Expert of Emission Reduction/Ecodesign/Sustainability/LCA/Energy Saving.

Ambient Intelligence as One of the Key ICT Factors for Energy Efficiency in Buildings

Antun Kerner[1], Dina Simunic[2] and Ramjee Prasad[3]

[1] Kerner Inc., 10000 Zagreb, Croatia;
e-mail: antun.kerner@ericsson.com
[2] Faculty of Electrical Engineering and Computing (FER), University of Zagreb,
Zagreb, Croatia; e-mail: dina.simunic@fer.hr
[3] Center for TeleInFrastruktur, Aalborg University, Aalborg, Denmark;
e-mail: prasad@es.aau.dk

Received 29 November 2012; Accepted 29 December 2012

Abstract

Ambient intelligence (AmI) is a multidisciplinary paradigm based on the Invisible Computer and Ubiquitous Computing. In AmI, technologies are deployed to make computers disappear in the background, while the human user moves into the foreground in complete control of the augmented environment. AmI is a user-centric paradigm, it supports a variety of artificial intelligence methods and works pervasively, no intrusively, and transparently to aid the user. AmI supports and promotes interdisciplinary research encompassing the technological, scientific and artistic fields creating a virtual support for embedded and distributed intelligence. In this paper authors are looking for possibilities to implement AmI as a way for saving energy in the construction industry, especially in buildings. AmI combines ICT and sensor technologies and enables monitoring, controlling and management of using energy in buildings and becomes key factor of energy efficiency in the construction industry.

Keywords: Ambient intelligence, energy efficency, sensor network, smart environment.

Journal of Green Engineering, Vol. 3, 219–243.

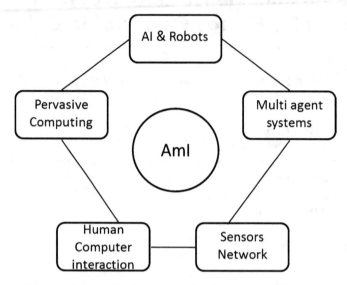

Figure 1 Relationship between AmI and several scientific areas.

1 Introduction

Ambient intelligence (AmI) started to be used as a term to describe "a digital environment that proactively, but sensibly, supports people in their daily lives" [1] type of developments about a decade ago and it has now been adopted as a term to refer to a multidisciplinary area which embraces a variety of pre-existing fields of computer science as well as engineering (see Figure 1).

Given the diversity of potential applications this relationship naturally extends to other areas of science like education, health and social care, entertainment, sports, transportation, construction etc. It may not be of interest to a user what kinds of sensors are embedded in the environment or what type of middleware architecture is deployed to connect them. Therefore, the main thrust of research in AmI should be integration of existing technologies rather than development of each elemental device.

On 16 February 2011, the General Assembly of the United Nations adopted a Resolution (A/RES/65/151) declaring the year 2012 to be the International Year of Sustainable Energy for All.

The European institutions, and especially the European Commission (EC), have acknowledged that the construction industry is a key industrial sector both in terms of employment and societal impact, and in terms of energy efficiency targets:

- The construction industry comprising construction, renovation, mainten-ance and demolition of buildings and infrastructures directly employs around 13.6 million persons and supports about 26 million jobs in Europe. It represents 6.2% of GDP. Demand in Europe is coming from private households, the business society and the public sector alike, the latter dominating demand for infrastructure building.
- In terms of energy efficiency, Europe has recognized "Energy Efficient Buildings" as a key industrial target – together with "Green Cars" and "Factories of the Future" – for improving the use of energy and de-creasing GHG emissions: Buildings (residential, public, commercial and industrial) account for approximately 40% of energy end-use in the EU, of which more than 50% is electrical power. The sector has significant untapped potential for cost-effective energy savings which, if realized, would mean an 11% reduction in total energy consumption in the EU by 2020.
- Energy Efficiency in buildings is estimated that the largest cost-effective energy savings potential lies in the commercial buildings (around 30%) and residential buildings (around 27%).
- Under the Directive on eco design of energy-using products (EuP) [2], implementing measures laying down requirements for energy and en-vironmental performance, are being enacted for ICT products used in the buildings and construction sector [3]. There is scope for ICT to contribute to further realizing this potential, through the application of building and energy management systems, smart metering technologies, solid-state lighting and lighting control systems, intelligent sensors and optimization software. In view of the contribution to energy performance of many different factors, including materials and in Europe techno-logies, and the various potential trade-offs among them, developing a systemic understanding of the energy performance of a building is highly desirable.
- The estimated envelope for the proposed Energy Efficiency Buildings Private-Public Partnership (EEB PPP) is 1 billion Euros. It is worth noticing that ICT is referred to as follows: "Research on ICT for energy efficiency in buildings, such as design and simulation tools, interoperab-ility/standards, building management systems, smart metering and user awareness tools".
- The proposed recast of the Directive on the Energy Performance of Buildings (EPBD) [5] introduces a general framework for a methodo-logy to calculate the energy performance of buildings. Implementation

of EPBD will yield a large amount of information on the makeup of building stock across Europe. Such information provides a useful baseline for the buildings and construction sector, as well as policy-makers. It also opens up opportunities for a collaboration framework between the ICT and buildings and construction sectors that is to be established with a view to exploiting opportunities for the development of software applications for the purpose of compliance with the EPBD.

- It is therefore a strategy acknowledged by the EC that the ICT sector is to be invited to work together with the buildings and construction sector to identify areas where the impact and cost-effectiveness of ICTs can be maximized, and to specify requirements. Ambient Intelligence is able to promote interoperability between auditing tools, and building and energy management systems, with a view to developing a systemic understanding of a building's energy performance. Then AmI could be applied for collection, aggregation and comparative analyses to support benchmarking and energy efficiency policy evaluation. AmI has an important role to play in reducing the energy intensity and therefore increasing the energy efficiency of the building where it has been implemented, in other words, in reducing emissions and contributing to sustainable growth. AmI is expected to generate a deep impact in the energy efficiency of buildings of tomorrow.

2 Current Situation in Croatia in the Construction Industry

Based on the alignment of Croatian legislation with EU legislation, as well as transfer of Directive 2002/91/EC on energy efficiency and Directive 2006/32/EC on end-use energy and energy services, the institutions responsible for implementing this program are established in the Republic of Croatia. These institutions are: Ministry Economy, Labor and Entrepreneurship (MELE) and the Ministry of Environmental Protection, Physical Planning and Construction (MEPPPC). Croatian Government adopted at the 14th session held 10th of April 2008. Action Plan for Implementation European Directive on Energy Performance of Buildings in the Croatian legislation.

Both ministries aligned their legislations with the requirements of Directive 2002/91/EC, the Ministry of Economy, Labor and Entrepreneurship adopted the law on efficient use of energy for end user consumption (Official Gazette (NN) 152/08), and the Ministry of Environmental Protection, passed the Physical Planning and Construction Act (NN 76/07, NN 38/09), of which in particular Articles 3, 4, 5, 6, 7 and 10 regulate the issues specified.

Additionally, the MEPPPC created:

- Technical regulation on the rational use of energy and thermal protection of buildings (NN 110/08, NN 89/09).
- Technical regulation on heating and cooling buildings (NN 110/08).
- Regulation on Energy Certification of Buildings (NN 113/08, NN 91/09, NN 36/10).
- Methodology for conducting energy audits of buildings (the decision of the Minister on 10 June 2009).
- Rules on conditions and criteria for persons who perform energy audits and energy certification buildings (NN 113/08, NN 89/09).
- Technical regulation on ventilation, partial air conditioning and Building Act (NN 03/07).
- Technical regulation for windows and doors (NN 69/06).
- Technical regulation for chimneys in buildings (NN 03/07).
- Regulation on Energy Certification of Buildings (Official Gazette 113/08, NN 91/09, NN36/10).
- The methodology of conducting energy audits of buildings (decision of 10 June 2009).
- Rules on conditions and criteria for persons who perform energy audits and energy certification of buildings (NN 113/08, NN 89/09).

Transfer of the Directive into the laws governing the energy and energy efficiency under the jurisdiction of the Ministry of Economy, Labour and Entrepreneurship, which adopted:

- Law on efficient use of energy in final consumption (NN 125/08)
- Law on Construction Products (NN 86/08)
- Law on energy efficiency in final energy consumption (NN 152/08)

The law on energy efficiency in the consumption of energy efficiency shall be the area of efficient energy use in the final. This law provides for the adoption of programs and plans for improving energy efficiency and their implementation, as well as energy efficiency measures. It also specifically regulates the activity of energy services and energy audits, and obligations of public sector undertaking and large energy consumers and consumer rights on the application of energy efficiency measures. It should be noted that this Act does not apply to energy efficiency in energy production and energy transformation, transmission and distribution of energy, and the conditions for conducting energy audits for the issuance of building energy certification, energy audits or to the boiler for heating and air conditioning systems in the building relating to special regulations in the field of construction. The purpose of this law is to achieve the goals of sustainable energy development in the following areas:

- Reducing negative environmental impacts from the energy sector.
- Improve security of energy supply.
- Meeting the needs of energy consumers.
- Fulfillment of international obligations in the Croatian area of reducing emissions of greenhouse gases and encouragement of energy efficiency in the sectors of final energy consumption.

The Croatian Standards Institute as a national standards body has adopted, through technical committees covering the area of energy efficiency in buildings, the majority of international and European standards as Croatian standards to which reference is made in the application of the rules on the energy certification of buildings.

3 Energy Performance of Building

By definition, the energy performance of buildings is the amount actually consumed or estimated consumption for different needs in accordance with the assumed (standardized) using the building and includes heating energy, domestic hot water heating, cooling, ventilation and lighting. The amount of energy is calculated taking into account insulation, technical and insulating properties, exposure solar light, the influence of neighboring structures, own-energy generation and other factors, including indoor climate, which affects the energy requirements. Accordingly, a general methodology is established for defining the energy performance of buildings, including:

- Thermal performance of buildings,
- Air,
- Heating,
- Installation of hot water,
- Ventilation,
- Air Conditioning,
- Lighting,
- Position and orientation of buildings,
- Passive solar systems,
- Protection from excessive solar radiation,

- Natural ventilation, and
- Indoor climate conditions.

The previously described methodology should take into account the positive impact of the following factors:

- Active solar systems and other systems for heating and electricity production based on renewable energy sources,
- Heat and electricity cogeneration,
- District heating and cooling, and
- Natural lighting.

Residential and non-residential buildings are divided into the following classes of energy:

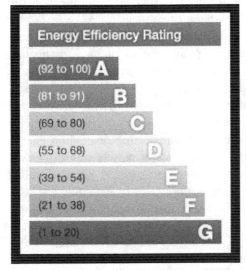

Class of Energy	QH, nd, ref Specific annual energy use for heating in the reference climatic data in kWh/(m^2)
A+	≤ 15
A	≤ 25
B	≤ 50
C	≤ 100
D	≤ 150
E	≤ 200
F	≤ 250
G	> 250

4 Rational Use of Energy and Thermal Protection in Buildings

According to the degree of complexity of the technical regulations on the rational use of energy and thermal protection in buildings (NN 110/08, 89/09) varies:

(a) Buildings with simple technical systems: residential or non-residential buildings without heating, cooling and ventilation, and with individual systems for domestic hot water, and buildings with individual and central heat source for heating without special heat recovery systems, with distribution thermal energy and the central or alternative systems for domestic hot water without using alternative systems, and individual cooling units, ventilation without heat recovery and limiting noise in ventilation systems without additional air treatment.

(b) Buildings with a complex technical system: residential or non-residential buildings with a central plant sources of heat for heating or cooling buildings with central hot water, with systems measurement and distribution of heating and cooling energy, central cooling systems, ventilation systems and air. conditioning systems with heat recovery and limit noise, air and additional treatment. It is also considered complex technical system and the building used for heating or cooling energy derived from alternative energy supply system, remote central heating or cooling, refrigeration systems, ventilation devices with controlled heating and cooling air and air-conditioning devices, including the associated cooling devices and other buildings that are not considered simple technical system.

Alternative energy supply systems are:

- Decentralized energy supply systems based on renewable energy,
- Cogeneration and trigeneration,
- District or block heating and cooling,
- Heat pumps,
- Condensing boilers and low-temperature, and
- Other systems with heat recovery.

Technical requirements for the rational use of energy and thermal protection in buildings are regulated by:

- Maximum permitted annual heating energy per unit of useful floor area of the building, or per unit volume of the heated part of the building.

- The maximum transmission coefficient of heat loss per unit of surface area of the heated part of the building.
- Preventing overheating of a building due to the effects of solar radiation during summer
- Air permeability limits the building envelope (for the time period when people stay required number of changes of air $n = 0.5 \, h^{-1}$, and while people do not stay $n = 0.2 \, h^{-1}$).
- The maximum thermal transmittance of certain parts of the building envelope.
- Reducing the impact of thermal bridges.
- The maximum allowed by condensation of water vapor inside of the construction of building.
- Prevent surface condensation.
- Placement of radiators.
- Limitation of coefficient of passing the heating thermal heating in panel.
- Mounting elements for the regulation of heat.
- Technical measures for the elements of heat distribution in buildings.
- Demands on the system of forced ventilation or air conditioning.
- Return to heat incoming air to ensure the building is ventilated with a number of mechanical devices air change $n > 0.7 \, h-1$ and the air flow $> 2500 \, m^3/h$.
- Prohibit the use of heating systems based on electric resistance from the year 2015 on. Until then, for the new building values QH, ndx1, 3 must be less than the allowable.
- Dynamic thermal characteristics of building elements of the building.
- Mandatory central plant for hot water in new buildings with more than three residential units, except for buildings with a connection to district heating, gas-heated buildings, houses in a row or if the annual consumption of energy is less than $25 \, kWh/(m^2 a)$.
- Separate calculation of energy performance of buildings if the design temperature varies by more of 4°C, for different purpose, different thermo system or is there a difference in the mode of use thermal engineering systems.

5 Primary Energy End-Use Splits

Most of the energy used in a home goes towards conditioning the space, which is often more affected by the size of the house than the number of

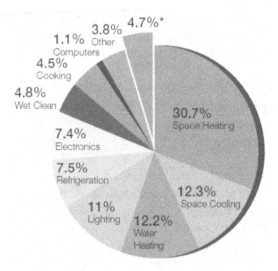

Figure 2 Residential primary energy end-use.

occupants. Heating, cooling, and lighting are still the largest single energy end-uses in a home, despite increased energy efficiency of this equipment.

How energy is used in a commercial building has a large effect on energy efficiency strategies. The most important energy end-use across the stock of commercial buildings is lighting, accounting for fully one-quarter of total primary energy use. Heating and cooling are next in importance, each at about one-seventh of the total. Equal in magnitude – although not well defined by the Energy Information Administration – is a catch-all category of "other uses" such as service station equipment, ATM machines, medical equipment, and telecommunications equipment. Water heating, ventilation, and non-PC office equipment are each around 6% of the total, followed by refrigeration, computer use, and cooking.

Balancing the risks of human health impacts and building materials damage with the benefits of energy savings and sound level reduction is a challenging matter. Decisions made will depend largely on the microclimate, state of existing insulation efficacy and budget available for insulation. Methods of improving on achieving balance of risks versus benefits may consist of the following structural changes in insulation decision making: aggressive promulgation of information about health risks that may arise from building insulation to homeowners, commercial property owners, building managers and local building officials; elimination of government involvement in insu-

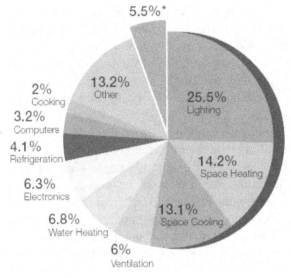

Figure 3 Commercial primary energy end-use.

lation programs, in favor of state and local programs, where knowledge of local meteorology and radon risks may be known better; removal of certain arbitrary insulation requirements on building skins, that may not have been developed with comprehensive understanding of health risks; and pursuit of research into improved methods of insulation that may allow a prescribed amount of air exchange through the building skin, and that might avoid use of toxic materials.

6 Energy Efficiency and Renewable Energy in Buildings

Virtually every part of a building's structure – from its placement and design to the appliances it contains affects its energy consumption. Climate-responsive architecture and whole building design consider the building's surroundings and local climate in order to construct energy-efficient buildings.

Proposed priorities by the EU:

1. Energy Performance Certification as a driver for step-by-step renovation: capturing the market.
   Actions bringing about increased uptake of the recommendations for energy efficiency and renewable energies of energy performance certi-

ficates. The actions should result in increased demand on the market for step-by-step renovation. This could include actions related to financing, resolving the owner/tenant dilemma, engaging consumers in relation to the significance of the recommendations made on buildings certification issued in line with the Directive on the energy performance of buildings (EPBD); bringing the industry together to develop one-stop shop solutions, etc.

2. Nearly Zero-Energy Buildings: transforming the existing building stock. Actions resulting in accelerated rates of refurbishment of existing buildings into Nearly Zero-Energy Buildings. This could include actions assisting the public sector on going beyond the proposed 3% renovation target, supporting the private sector, bringing together industry elements to provide packaged solutions, promoting frontrunners, etc.

3. Building as designed: quality and compliance in construction. Actions resulting in improved quality in construction and compliance to building codes in support of Article 10 of the recast EPBD (Directive 2010/31/EU) and Article 13 of the RES Directive (Directive 2009/28/EC). This could include market observatories, quality seals, etc, resulting in increased consumer confidence and demand for high quality construction in new buildings and renovations. In addition, establishment of robust benchmarks and knowledge of the actual performance of early renovations and installations will be a pre- requisite.

As previously mentioned in this article we use AmI as contribution in theory which demonstrates that energy consumption can be lowered by effective control and management. Efficient use of energy is dependent of using energy efficient materials, devices, systems and technologies that are available on the market, to reduce consumption energy to achieve the same effect (heating, cooling, lighting, cooking process, washing, etc.). The US Environmental Protection Agency and the US Department of Energy support the Energy Star approach based on the PDCA (Plan Do Check Act) concept (see Figure 4). Energy efficiency is the fastest, cheapest, and largest untapped solution for saving energy, saving money, and preventing greenhouse gas emissions. Through Energy Star, EPA has helped thousands of businesses and organizations tap these savings in the places where people live, work, play, and learn.

According to the US Department of Energy's Energy Information Administration, commercial buildings alone account for 18% of the total energy consumption. A major overhaul at the iconic Empire State Building helped

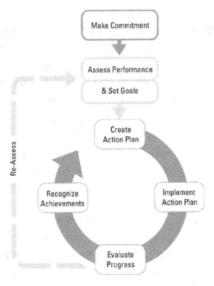

Figure 4 Energy Star management and control aproach.

raise the profile of energy efficiency. This project, which included replacing 6,500 windows, adding insulation, upgrading lighting, and installing a digital wireless monitoring system is powering a 38% annual energy reduction and $4.4 million in annual savings.

7 AmI Applications

AmI is a developing approach of using different ICT and sensors technologies which consists of applications which can in specific cases contribute in energy efficiency.

7.1 Ambient Intelligence in the Buildings and City

In the US and Canada, a large number of community networks supported by grass-root activities appeared in the early 1990s. In Europe, more than one hundred digital cities have been tried, often supported by local or central governments and the EU in the name of local digitalization. Asian countries have actively adopted the latest information technologies as a part of national initiatives. In the past 15 years since the first stage of digital cities, the development of the original digital cities has leveled off or stabilized. In spite of that, by looking back at the trajectory of the development of digital cities, it is clear that digital environments in cities have often benefited from the previous activities on various regional information technology spaces.

User-oriented design, privacy preservation, personalization and social intelligence are some of the fundamental concepts the area has to approach efficiently in order to encourage acceptance from the masses.

The basic idea of ambient intelligence is to connect and integrate technological devices into our environment such that the technology itself disappears from sight and only the user interface remains. Representative technologies used in ambient intelligent environments are collaborative machine assistants (CMAs), for example, in the form of a virtual agent or robot.

It is expected to develop a series of recommendations for ambient intelligence: general; service aspects; network aspects; human and ambient interface and protocol; physical layer on the wireless path; speech and video coding specification; terminal adaptor for human: ambient to ambient interface; network inter working; service inter working; equipment and type approval specifications; operation and maintenance.

Applications are an important part of AmI architecture and to enable communication between sensors implemented in building. Middleware architecture is used (see Figure 5).

AmI has the characteristics of a smart environment and it has an allocentric view. It needs location maintenance and it is based on information availability. From a sensors' point of view it allows the sensors' independence and it defines sensory equipment. Our assumption in this article is that using network of sensors as input for AmI application is possible to achieve efficient control and management of building where network of sensors and AmI application is implemented. In this way AmI enables energy saving.

7.2 System for Setting Ground Rules and Submission in Building

Important element in management of building is an application on a user device that takes data from the system setup and submission in the space/environment.

The assumption is that the user will have a positive impact on using of space if there is information on what is expected of him. Simple examples are temperature and lighting, more complex examples can be shortcut definition, noise, etc. The prototype would be a monitor at the entrance of a

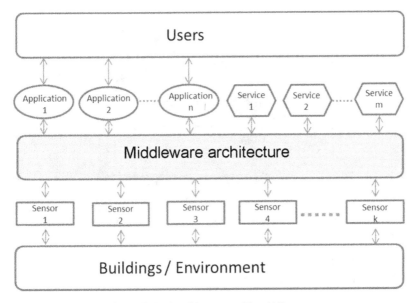

Figure 5 AmI architecture with middleware.

particular zone, the message about the availability of basic rules for dealing with space/environment, data exchange with a person who enters the space/ambient. Download the data may be arbitrary or optional. Once you step into the Ambient and the user share data, the user's device may show recommendations, warnings about deviations/differences. Users can decide whether to turn off the light because it is the intensity of illumination greater than defined in the downloaded recommendation, the temperature deviates from the level recommended by the host, etc. New buildings would be equipped with such systems, and the extension into existing buildings would be easy. This would be drive mechanisms for the introduction house rules.

7.3 Sensor Networks Applications

Sensor networks may consist of many different types of sensors such as seismic, low sampling rate magnetic, thermal, visual, infrared, acoustic and radar, which are able to monitor a wide variety of ambient conditions that include the following:

- Temperature,
- Humidity,
- Vehicular movement,

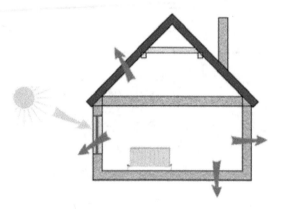

Figure 6 Positions of sensors related with loses of energy.

- Lightning condition,
- Pressure,
- Soil makeup,
- Noise level,
- The presence or absence of certain kinds of objects,
- Mechanical stress levels on attached objects, and
- The current characteristics such as direction and size of an object.

Sensor nodes can be used for continuous sensing, event detection, event ID, location sensing, and local control of actuators. The concept of micro-sensing and wireless connection of these nodes promises many new application areas.

AmI consists of different integrated applications and in combination with using new technologies and building materials properly has a significant impact on energy savings and energy efficiency.

We can note the following systems and characteristics:

1. HEMS – Home/Building Energy Management System

   - Autonomous control of optimum A/C by season and operation mode.
   - Optimization of electricity usage with "CO_2 minimization mode" and "energy cost minimization mode" as well as "visualization" of CO_2 emissions and energy cost.
   - System to minimize the effects of outages

2. . V2H – Vehicle to Home System

   - Effective use of electric vehicle battery.

- Electricity can be supplied from vehicle to home on days when vehicle is not to be used or to minimize effects of outages.
- Vehicle battery can by remotely controlled by mobile phone.

3. Home-use sustainable energy
   - Natural energy, which is easily affected by weather conditions, to be stored in batteries.
   - Stable ecological power available from batteries.

4. Natural ventilation
   - Linked to HEMS, ventilation system operates autonomously sensing weather conditions.
   - When operating, opening and closing of vents is automatically controlled by wind pressure to adjust room temperature.

5. Walls and rooftop isolation and greening
   - Room temperature rise limited, load on A/C cut.
   - Direct sunlight blocked by greening of walls and roof.
   - Comfort and energy saving with effective placement of plants.

6. Electronic window management
   - Electronic film curtain on window can be made opaque with a switch.
   - Power saving by allowing in natural light, and privacy.

7. Home-use fuel cell
   - Total electrical and heat efficiency and durability of fuel cell.

8. Home server
   - Platform to control systems such as HEMS and V2H based on Web technology.
   - Control of home appliances by mobile phone.
   - Home security, control and monitoring of home appliances.
   - Comfortable and convenient living through ICT with multi-use server and wireless sensor network.

9. Smart outlets
   - Advanced multifunction power outlets controllable via the Web.
   - Working in combination with HEMS to monitor and control each home appliance.
   - Priority of each outlet can be set to cut power in case of outage.

10. Compact, advanced power storage device
    - Power is stored at night for home-appliance standby use.
    - In case of outage or instantaneous voltage drop, stored power is immediately discharged by linkage to HEMS to minimize the effect (High – e.g. Server; Medium – e.g. TV; Low – e.g. A/C).

11. Breaker with indicator light
    - When a breaker trips, even in darkness, the switch can be located with certainty,

12. Induction heating
    - Application of IH technology.
    - No heat source in the body of the iron for instance means there is no danger of burning.
    - Preheating unnecessary.

13. Home health management
    - Daily home health check in the bathroom.
    - Automatic measurements taken in the bathroom include for example urinary sugar and blood pressure.
    - Results stored and analyzed in home server.

Before we choose the technology to implement all necessary sensors to enable monitoring and management of using energy, we reviewed several technologies used in ambient intelligence applications. The first one was developed in 1975, when a company in Scotland developed X10. X10 allows compatible products to talk to each other over the already existing electrical wires of a home. All the appliances and devices are receivers, and the means of controlling the system, such as remote controls or keypads, are transmitters. If you want to turn off a light in another room, the transmitter will issue a message in numerical code that includes the following:

- An alert to the system that it is issuing a command.
- An identifying unit number for the device that should receive the command.
- A code that contains the actual command, such as "turn off".

All of this is designed to happen in less than a second, but X10 does have some limitations. Instead of going through the power lines, some systems use radio waves to communicate. However, building automation networks don't need all features of a WiFi network because automation commands are short messages. The two most prominent radio networks in building automation are

ZigBee and Z-Wave. Both of these technologies are mesh networks, meaning there is more than one way for the message to get to its destination.

Z-Wave uses a Source Routing Algorithm to determine the fastest route for messages. Each Z-Wave device is embedded with a code, and when the device is plugged into the system, the network controller recognizes the code, determines its location and adds it to the network. When a command comes through, the controller uses the algorithm to determine how the message should be sent. Because this routing can take up a lot of memory on a network, Z-Wave has developed a hierarchy between devices: Some controllers initiate messages, and some are "slaves", which means they can only carry and respond to messages.

ZigBee's name illustrates the mesh networking concept because messages from the transmitter zigzag like bees, looking for the best path to the receiver. While Z-Wave uses a proprietary technology for operating its system, Zig-Bee's platform is based on the standard set by the Institute for Electrical and Electronics Engineers (IEEE) for wireless personal networks. This means any company can build a ZigBee-compatible product without paying licensing fees for the technology behind it, which may eventually give ZigBee an advantage in the marketplace. Like Z-Wave, ZigBee has fully functional devices (or those that route the message) and reduced function devices (or those that do not).

Using a wireless network provides more flexibility for placing devices, but like electrical lines, they might have interference. Some solutions offer a way for home/building network to communicate over both electrical wires and radio waves, making it a dual mesh network. If the message is not getting through on one platform, it will try the other. Instead of routing the message, a device will broadcast the message, and all devices pick up the message and broadcast it until the command is performed. The devices act like peers, as opposed to one serving as an instigator and another as a receptor. This means that the more devices that are installed on a network, the stronger the message will be.

There are currently several million km of existing data cable installed within commercial buildings. Wireless building solutions embody the prevailing goal of sustainable buildings: Reduce, Reuse and Recycle. Fewer wired and more wireless solutions mean less disposed cabling will end up in landfills and less waste will be burned and emitted as toxic gases into the environment.

System Retrofits: Wireless solutions are ideally suited for existing spaces because they eliminate the need to remove floors, walls or ceilings to access

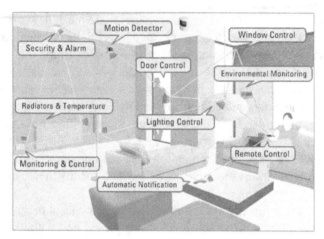

Figure 7 Different sensors distributed in building constitute ambient intelligence and contribute in energy efficiency.

control products. People or processes no longer need to relocate while upgrades are under way, allowing continued access to labs, sensitive storage, health facilities and critical process areas.

Mobile Applications: The ability to monitor the home via smart devices is a trend, since smart phones and tablets are changing user habits and gaining ground as the primary interface of Home Automation (HA) systems. HA systems now offer apps to let customers use mobile devices to control them.

8 AmI Applications – Example

Connect the gateway to an open network port of your existing internet router. Using this option, the data from one re more ECM-1240 devices is sent directly to a professional data hosting center. (Requirement is an existing high speed internet connection and internet router with a spare Ethernet port (standard RJ-45 connection).)

Some of the many options include real-time and historical graphical and tabular data are viewable over the internet. Charts may be viewed using iPhone or similar devices. Energy conservation performance report cards automatically emailed. Email or text alerts of potential issues such as: fridge not running, sump-pump not cycling, peak demand about to be exceeded, system down/power outage, to name a few. Indoor, outdoor temperature options Net metering capable.

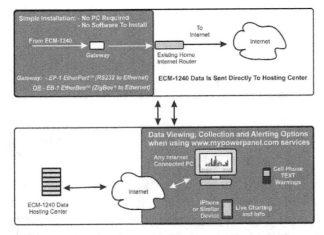

Figure 8 Sensors distributed in building enables measure power purchased from power company and power used by individual house loads.

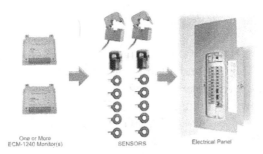

Figure 9 Sensors distributed in building enables measure power purchased from power company and power used by individual house loads.

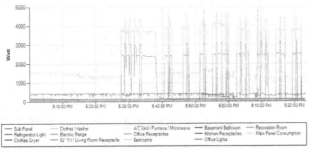

Figure 10 The power used by various loads are charted and may be viewed "live".

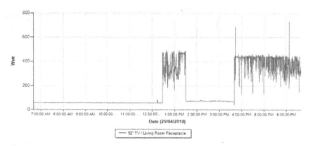

Figure 11 Example: "Living Room TV" – load profile makes the ON-OFF times obvious.

The previously mentioned examples illustrate that the bottom up approach enables collecting data at different levels and according to that information ambient intelligence is able to contribute in energy efficiency by automatic tuning of energy consumption or by proper suggestions to the facility manager or even to the end user of particular space in a specific building.

9 Conclusion

In most countries, buildings are the largest driver for both energy use and CO_2 emissions. The 160 million buildings of the EU, for example, are estimated to use over 40% of Europe's energy and to drive over 40% of its carbon dioxide emissions. According to the US Energy Information Administration (EIA), the share of energy and GHG emissions associated with buildings is even larger in the US, with 48% of the total. Top down methodology is driven by EU Commission and Governments. AmI applications enable bottom up methodology for increase of energy efficiency of buildings. A simple example shows that implementation of wireless sensors combined with monitoring tool and wireless power switch can, in the shape of different AmI applications, contribute in energy efficiency.

Information is power: At least a third of the carbon savings in the residential and commercial sectors come from behavioural changes. Information about the amount of Energy a householder is using is essential to help consumers to reduce consumption and increase efficiency. Smart Meters are being developed to enable the utilities to take remote meter readings – essential for providing customers with regular and accurate power bills. Energy efficiency on its own is insufficient. This is because it is usually a relative, not absolute figure. For instance, with refrigerators, the energy efficiency index comes from the relationship between total electricity use over a year and the volume of the interior of the refrigerator. As a result, the manufacturers have been producing more energy efficient refrigerators, that are both larger and, in absolute terms, consuming more electricity. The same situation occurs with housing, e.g. washing machines. Our work shows that the availability of information about using energy by connecting buildings to the Internet can contribute to changing behavior of building/home users and owners. Some additional research should define how many buildings and users constitute a representative pattern to confirm that statement.

References

[1] J.C. Augusto. Ambient Intelligence: The Confluence of Ubiquitous/Pervasive Computing and Artificial Intelligence, pp. 213–234, Springer, London, 2007.

[2] Action Plan for Energy Efficiency: Realising the Potential. Commission of the European Communities, Brussels, 19 October 2006.

[3] Directive 2009/125/EC of the European Parliament and of the Council. *Official Journal of the European Union*, 31 October 2009.

[4] COM (2009) 111 final. Communication from the European Commission ... on mobilising information and communication technologies to facilitate the transition to an energy-efficient, low-carbon economy. 2009. Available from cordis.europa.eu/fp7/ict/sustainable-growth/key-docs_en.html, ec.europa.eu/information_society/activities/sustainable_growth/docs/com_2009_111/com2009-111-en.pdf.

[5] COM (2008) 800 final. A European Economic Recovery Plan. Communication from the Commission to the European council. Brussels, 26 November 2008. Available from ec.europa.eu/commission_barroso/president/pdf/Comm_20081126.pdf.

[6] Directive on the Energy Performance of Buildings (EPBD). European Parliament and the Council. December 2002. Available from www.buildingsplatform.org.

[7] R. Prasad, S. Ohmori, and D. Simunic (Eds.). Towards Green ICT. River Publishers, Aalborg, 2010.

[8] M. Young. The Technical Writer's Handbook. University Science, Mill Valley, CA, 1989.

[9] D. Estrin, R. Govindan, J. Heidemann, and S. Kumar. Next century challenges: Scalable coordination in sensor networks. In Proceedings of ACM MobiCom'99, Washington, DC, , pp. 263–270, 1999.

[10] S. Tobgay, R.L. Olsen, and R. Prasad. Adaptive information access on multiple applications support wireless sensor network. In Proceedings CSNDSP'12, pp. 1–4, 2012.

[11] J. Spohrer, P. Maglio, J. Bailey, and D. Gruhl. Intelligent computing everywhere. Steps toward a science of service systems. IEEE Computer, 40(1):71–77, 2007.

Biographies

Antun Kerner is an independent researcher. He graduated at the University of Zagreb, Faculty of Electrical Engineering in Zagreb, Croatia. He worked on different communications and telecommunications projects. He is the author or co-author of approximately ten publications.

Dina Simunic is a full professor at the University of Zagreb, Faculty of Electrical Engineering and Computing in Zagreb, Croatia. She graduated in 1995 from the University of Technology in Graz, Austria. In 1997 she was a visiting professor in Wandel & Goltermann Research Laboratory in Germany, as well as in Motorola Inc., Florida Corporate Electromagnetics Laboratory, USA, where she worked on measurement techniques, later on applied in IEEE Standard. In 2003 she was a collaborator of USA FDA on scientific project of medical interference. Dr. Simunic is a IEEE Senior Member, and acts as a reviewer of *IEEE Transactions on Microwave Theory and Techniques, Biomedical Engineering and Bioelectromagnetics, JOSE*, and as a reviewer of many papers on various scientific conferences (e.g., IEEE on Electromagnetic Compatibility). She was a reviewer of Belgian and Dutch Government scientific projects, of the EU FP programs, as well as of COST-ICT and COST-TDP actions. She is author or co-author of approximately 100 publications in various journals and books, as well as her student text for wireless communications, entitled: *Microwave Communications Basics*. She is co-editor of the book *Towards Green ICT*, published in 2010. She is also editor-in-chief of the *Journal of Green Engineering*. Her research work comprises electromagnetic fields dosimetry, wireless communications theory and its various applications (e.g., in intelligent transport systems, body area networks, crisis management, security, green communications). She serves as Chair of the "Standards in Telecommunications" at Croatian Standardization Institute. She servers as a member of the core group of Erasmus Mundus "Mobility for Life".

Ramjee Prasad (R) is currently the Director of the Center for TeleInfrastruktur (CTIF) at Aalborg University (AAU), Denmark and Professor, Wireless Information Multimedia Communication Chair. He is the Founding Chairman of the Global ICT Standardisation Forum for India (GISFI: www.gisfi.org) established in 2009. GISFI has the purpose of increasing the collaboration between European, Indian, Japanese, North-American, and other worldwide standardization activities in the area of Information and

Communication Technology (ICT) and related application areas. He was the Founding Chairman of the HERMES Partnership – a network of leading independent European research centres established in 1997, of which he is now the Honorary Chair.

Ramjee Prasad is the founding editor-in-chief of the Springer *International Journal on Wireless Personal Communications*. He is a member of the editorial board of several other renowned international journals, including those of River Publishers. He is a member of the Steering, Advisory, and Technical Program committees of many renowned annual international conferences, including Wireless Personal Multimedia Communications Symposium (WPMC) and Wireless VITAE. He is a Fellow of the Institute of Electrical and Electronic Engineers (IEEE), USA, the Institution of Electronics and Telecommunications Engineers (IETE), India, the Institution of Engineering and Technology (IET), UK, and a member of the Netherlands Electronics and Radio Society (NERG) and the Danish Engineering Society (IDA). He is also a Knight ("Ridder") of the Order of Dannebrog (2010), a distinguishment awarded by the Queen of Denmark.

Online Manuscript Submission

The link for submission is: www.riverpublishers.com/journal

Authors and reviewers can easily set up an account and log in to submit or review papers.

Submission formats for manuscripts: LaTeX, Word, WordPerfect, RTF, TXT.
Submission formats for figures: EPS, TIFF, GIF, JPEG, PPT and Postscript.

LaTeX

For submission in LaTeX, River Publishers has developed a River stylefile, which can be downloaded from http://riverpublishers.com/river_publishers/authors.php

Guidelines for Manuscripts

Please use the Authors' Guidelines for the preparation of manuscripts, which can be downloaded from http://riverpublishers.com/river_publishers/authors.php

In case of difficulties while submitting or other inquiries, please get in touch with us by clicking **CONTACT** on the journal's site or sending an e-mail to: info@riverpublishers.com

www.ingramcontent.com/pod-product-compliance
Lightning Source LLC
LaVergne TN
LVHW012331060326
832902LV00011B/1833